D0149571

THE PLAYFUL WORLD

The PLAYFUL World

HOW TECHNOLOGY IS TRANSFORMING OUR IMAGINATION

Mark Pesce

BALLANTINE BOOKS | NEW YORK

A Ballantine Book
The Ballantine Publishing Group

www.randomhouse.com/BB/

Grateful acknowledgment is made to the following for permission to reprint previously published material:
California Institute of Technology: excerpts from "There's Plenty of Room at the Bottom" by Richard P. Feynman, *Engineering and Science,* Vol. XXIII, No. 5, February 1960, California Institute of Technology.
Doubleday: excerpts from *Engines of Creation* by Eric Drexler. Reprinted by permission of Doubleday, a division of Random House, Inc.
Scientific American: excerpt from "50, 100, 150 Years Ago" column from December 1949 issue of *Scientific American*. Reprinted by permission of *Scientific American*.
The National Enquirer. Inc.: "FIRE! Furby Saves Soap Star's Life" from *The National Enquirer*, March 16th, 1999. Reprinted by permission of *The National Enquirer*, Inc.
The Estate of R. Buckminster Fuller: excerpts from the writings and lectures of R. Buckminster Fuller. © Estate of R. Buckminster Fuller.

Library of Congress Card Number: 00-106984

ISBN 0-345-43943-0

Text design by Ann Gold

Manufactured in the United States of America

First Edition: October 2000

10 9 8 7 6 5 4 3 2 1

For Stafford Bey Redding O'Brien (1999-),
who inspired,
and Terence McKenna (1946-2000),
who provoked

The Aeon is a child at play with colored balls.

—Heracleitus

CONTENTS

THE PLAYFUL WORLD

INTRODUCTION

A child born on the first day of the new millennium will live an entire lifetime in a world undreamt of just a generation ago. As much as we might have tried to speculate upon the shape of things to come, the twenty-first century arrives just as unformed as a newborn. We know this much: some of our future dreams have failed to appear—space travel remains a luxury for the few wealthy governments of the world—while others, such as the revolutions in information and communications, have exceeded everyone's most passionate fantasies. Space has been replaced by cyberspace, rockets by routers.

As a child during the heady days of Project Apollo, I had astronaut wallpaper in my bedroom and breakout diagrams of the titanic Saturn V rocket tacked to my walls. My treasure, though, was a toy astronaut, complete with a complex array of cables to simulate space walks, which tangled up impossibly almost the moment I took it from its box. I didn't mind, not one bit, because the figurine could still be taken in hand and flown through the universe of my imagination. Space-crazy, like most small boys in the late 1960s, ready to blast off at a moment's notice to points unknown, I had littered my personal world with all of the objects that reflected my own belief in a future among the stars.

My parents, complicit in this enterprise from the very beginning and caught up in the last great dream of Camelot—to send a man to the Moon and return him safely—fed my curiosity in every way they knew how. Books, star charts, and excursions to the planetarium at Boston's famed Museum of Science reinforced my daydreaming, each filling my mind with visions of the revealed mystery of the cosmos. Even today I keep abreast of the latest findings from the scientists who study the universe, and my mind still leaps among the stars.

When a child enters the world, it knows nearly nothing of the universe beyond itself. With mouth, then eyes, and finally hands, it reaches out to discover the character of the surrounding world. Over the course of time, that child will discover its mother—the source of life—and, some time later, its father. But in the first days after birth, the child will be presented with rattles, mobiles, mirrors, and noisy stuffed animals that will become its constant companions. Our children, in nearly every imaginable situation, are accompanied by toys.

It has been this way for a very long time. We can trace the prehistoric sharpened stick—undoubtedly the first tool—to the sticks children still love to play with today. Over the 5,500 years of recorded history, forward from Sumer and Egypt, toys have a presence both charming and enlightening, for we have learned that toys not only help to form the imaginations of our children, but also reflect the cultural imagination back upon us. The ancient Maya, who thrived across Mesoamerica thirteen centuries ago, never developed the wheel for transportation—already in use for some 7,000 years in Mesopotamia—yet employed it in toys. The Mayan worldview—based in circles and cycles of sky and earth—brought them the wheel as a toy, a pocket universe reflecting the structure of the whole cosmos.

All of our toys, for all of known time, perform the role of reducing the complex universe of human culture into forms that children can grasp. It is not that children are simple, unable to apprehend the complex relationships that form cultures. Rather, toys help the child to guide itself *into* culture, serving as playgrounds where rehearsals for reality can proceed without constraint or self-consciousness. When I played with toy astronauts, I awakened an appetite for science that has nourished my own career, just as thirty thousand years ago, another boy played with a spear and practiced the rituals of the hunt.

These points have been made before, but have gained unusual currency over the last few years, as the character of our toys has begun to change, reflecting a new imagining of ourselves and the world we live in. Somewhere in the time between Project Apollo and the Mars Pathfinder we learned how to make the world react to our presence within it, sprinkling some of our intelligence into the universe at large in much the same way a chef seasons a fine sauce. Our toys, touched by fairy dust, have come alive, like Pinocchio; some—like the incredibly popular Furby—simulate ever-more-realistic personalities.

In times past, children would create fantasies for themselves, conversing with their stuffed animals and dolls. Over a hundred years ago, Thomas Edison adapted his wax-cylinder phonograph to create the first popular "talking" toys, but until 1998, these mechanical contrivances were limited to only a handful of phrases.

Suddenly the toy talking to a child has its own language (Furbish), has the ability to compose simple sentences, and responds to a number of verbal and physical commands. Beyond this, it also has a will—if such a word can be applied to a toy—complete with requests for food and attention. Ignoring these requests exacts a price; if you forget to feed a Furby, it gets sick. Finally, Furbys *learn*. Their constant interactions with people make them more communicative, and each comes equipped with a simple infrared communications set (similar to a television and its remote control) so that one Furby can share its learning with another.

Although the Furby seems to have come from nowhere to capture the hearts of children worldwide, in reality it incorporates everything we already know about how the future will behave. The world reacts to us, interacts with us, at a growing level of intelligence and flexibility. A century ago people marveled at the power and control of the electric light, which turned night into day and ushered in a twenty-four-hour world. Today we and our children are amazed by a synthetic creature possessing a dim image of our own consciousness and announcing the advent of a playful world, where the gulf between wish and reality collapses to produce a new kind of creativity.

We give children toys to engage their imaginations, and in particular to translate their imaginings into tangible form. As a species we tend to regard the artifact over the fact: learning is nice, but doing is better. Kids will turn a sandpile into an empire or a stack of rotted wood into a tree fort. Lincoln Logs and Tinkertoys, Erector sets and Lego bricks all allow children to translate into reality something dreamed of only in the mind's eye. Those famous Danish blocks have been touched by intelligence, too. At the same time the Furby took off, Lego's Mindstorms began a similar climb into popularity. The brightly colored bricks have become components in a do-it-yourself robotic construction kit that children can easily learn to program into an infinite array of forms. The act of creation, which had meant working with fixed, static materials, suddenly encompasses the idea of the physical world as intrinsically manipulable, programmable, and mutable.

Lego Mindstorms doesn't stop there; created in the age of the World Wide Web, the company encourages youngsters to log on to a busy website and share their discoveries with others. Every child can learn from every other, and every new innovation in Mindstorms programming or design instantaneously becomes available everywhere, for the bricks themselves can talk to the Web. These toys define a world both intrinsically manipulable and infinitely connected across a haze of invisibly operating and ever-more-present digital networks.

Even as toys grow more flexible, more reflective of our increasing capabilities to shape the environment to suit ourselves entirely, they become a sort of magical laboratory for the exploration of possibilities entirely beyond our abilities. The hoopla associated with virtual reality in the early part of the 1990s has receded from view, but the technology behind it has become nearly ubiquitous. Hundreds of millions of homes around the world have incredibly powerful virtual reality systems, but in the disguise of video game consoles they have mostly been viewed by the adult population as throw-away devices, which merit little attention except concerning the violence presented in their programming.

Simulation, a sophisticated way of "making pretend," gave us code-breaking computers in the 1940s and won the Second World War; simulators helped astronauts fly to the Moon safely and help train jet pilots today. When simulation broke loose of its reality moorings in the first gusts of the high-technology hurricane, the world renamed it "virtual reality." VR came to culture as a sort of magical machinery for the creation of an infinity of worlds, however improbable exactly the kind of toy no child could resist.

In the autumn of 2000, when Sony Corporation releases its much-anticipated PlayStation 2, the machinery of infinite realities will be within the grasp of millions of children around the world. Unlike any videogame console released before it, the PlayStation 2 will have the power to create realistic imaginings of breathtaking clarity. Million-dollar computers—in 1999!—have only fractionally more power than the PlayStation 2, which will challenge our ideas about simulation by making it look at least as real as anything else seen on a television screen.

But the television cannot talk back, it can only broadcast. The PlayStation 2, designed to connect to the Internet, to connect *across* the Internet, becomes more than a toy; it becomes a window onto a wider world, a web of worlds, each more fantastic than the one preceding it. Its flexibility goes hand in hand with its virtuality; it could easily simulate a Furby (though you could not touch it) or Mindstorms (though you could not touch them), but it can create a million other interesting forms, if only for the eyes and ears—a spaceship for scouring the universe of ideas.

These three toys—Furby, Lego Mindstorms, PlayStation 2—can serve as points of departure for another voyage of exploration, a search for the world of our children's expectations. As much as a spear or wheel or astronaut figurine ever shaped a child's view of the world, these toys—because they react to us—tell us that our children will have a different view of the world, seeing it as potentially vital, intelligent, and infinitely transformable. The dead world of objects before intelligence and interactivity will not exist for them, and as they grow to adulthood, they will likely demand that the world remain as pliable as they remember from their youngest days.

Fortunately, we are ready for that challenge.

Our toys, writ large, echo profound revolutions in simulation, the science of materials, and digital communication. The technique of the Furby has been a hot topic of computer science for a dozen years; artificial life—simulation of the activity of living systems—has taught us a lot about how we learn and grow into intelligence. Computers, which just a decade ago seemed useful only for word processors and spreadsheets, are now employed as digital gardens, where the seeds of mind grow into utterly unpredictable forms. The rough outlines of such capabilities made their way into the Furby, but its successors will display ever-increasing levels of intelligence. The science of digital ecologies is only just beginning.

More than just simulations running in the mind of a computer, artificial life has worked its way into the real world, in a variety of robotic forms. From robotic "insects" to intelligences with faux human bodies, these robots learn from their continuous interactions with the environment, defining goals and changing strategies as they encounter the world. These machines have crossed an imaginary line from procedural to unpredictable, which delights their creators, who are in essence building the next generation of Furbys. Encountering a quirky, nonhuman, but thoroughly real intelligence is thrilling to both children and adults. In some way, it is life, and we instinctively recognize it.

Over the next twenty years we will endow ourselves with creative abilities beyond any we have ever known, because we have begun the full-scale exploration of the physical limits of our ability to manipulate the fine structures of the universe. Atoms have been known of for a hundred years, but until a decade ago they had to be considered in groups—chains, crystals, rings—in quantities so vast that even the smallest visible amount contained an unthinkable abundance. While chemists using components that look like so many Tinkertoys built models of molecules, it remained beyond their ability to place atom against atom individually and to build molecules from the ground up. Advances in equipment—such as the scanning tunneling microscope (STM)—opened the atomic world to view in a way that had only been imagined. More than this, the STM provided a finger to poke the universe's smallest components, even to knit them together.

For some years, researchers have been designing their own arrangements for atoms, drafts of simple mechanical structures such as switches, gears, and chains. From these basic mechanics, an immense array of machines can be built, including entire computers smaller than the smallest cell within a human body, computers that run a million times faster than those of the year 2000, communicate with the outside world, and can be programmed to build absolutely anything. Including other tiny computers.

Just as the creative world of children has become manipulable, programmable, and mutable, the entire fabric of the material world seems poised on the edge of a similar transformation. That, in essence, is the theme of this book, because where our children are already going, we look to follow.

Our relationship to the world of information is changing, because the hard-and-fast definitions of *world* and *information* have begun to collide, and the boundaries between them—which separate reality from imagination and idea from realization—have become ever more tenuous. In the era our children will inhabit, the world *is* information, and like information it can be stored, retrieved, processed, and portrayed in an endless abundance.

Hippocrates, the ancient father of medicine, said, "In all abundance there is lack." Abundant power to shape the world does not confer the wisdom to do so rightly; we will need to work from a knowledge and experience we do not yet possess. We need a playground where we can explore without fear of disaster. In short, we will need an imagination, a shared digital imagination, where we can experiment with the forms of the future.

Though virtual reality has lost its visibility, it has increasingly become a practical playground for the imagination of forms. Virtual reality has set its sights on physical reality, and—after a startling combination of an STM with a VR system in 1993—virtual reality has become the premier playground for the study of the impossibly small atomic systems dreamed of by researchers. Now we can see and touch the basic elements of the world, all in virtual reality. Soon it will be possible to assemble those tiny machines, test them, and send them into the world, but only because we have played with them first in a virtual sandbox.

A virtual world can magically invert scale, zooming from the single atom to the entire surface of the Earth. Already, satellites and ground systems meticulously and continuously scan the planet, then write their images across the World Wide Web. A few years ago, one VR project used a million dollars of computer power to create a detailed *live* vision of Earth; just a few months later something similar showed up on the Web. This was no more than a toy version of the megabuck production, but both pointed the way to another intersection of virtual reality and reality. If VR can grasp the very small, it can also put the whole world in our hands.

But virtual reality also belongs to the imagination, to our dreams, and while most of our VR dreams have been about tanks or airplanes or starships, some artists have found within it a vehicle to the soul. Just as the architecture of a great medieval cathedral can make the spirit soar, the spaces created in simulation can ring true within us, enveloping us in the ineffable. The atom, the world, and the dream—each of these virtualities can be created in the lab, but before this book reaches you, all of it will come to us through a toy in the living room, connected to a television and a digital network.

A tendency to overvalue the ends of technology has become one of the most persistent features of these heady times, but so much technology has been piling up for so long that we are now beginning to see how it transforms the way we think. We are *different* for using it. This qualitative change can be seen most clearly in the World Wide Web, which grew from a simple, if subtle, idea into a global unification of all human knowledge, and, perhaps, a catalogue of human experience. Confronted with a space of ideas that has grown well beyond the ability of any person to "know" it, we find ourselves navigators in a familiar but impossibly vast sea of facts, figures, and fiction. Every individual who has become a web surfer has changed the way he thinks and the way he uses knowledge. Every business, as it encounters the Web, changes completely.

A child born at the dawn of the third millennium will never know a world without the Web. A phenomenon that represents a serious midcourse correction for most of us—and our institutions—will be completely natural to them, and as pervasive as the air they breathe. It will give them a different view of the world, as television did to the children of two generations ago. But because the Web *reacts*, it will also give them the means to change the world to fit their own desires. Already we can see how we can reach through the Web to order a book, move a camera, or plant a garden. As the technologies of tiny machines and their "grown" robot brethren achieve their own reality, they will become the steadfast servants of a generation who can think locally yet act globally, a near perfect inversion of the generations raised on television, who saw at a distance but could only act on the ground beneath their feet.

In the evolving relationship between imagination and reality, toys show us how we teach the ways of this new world to our children. Their toys tell them everything they need to know about where they are going, providing them the opportunity to develop a mind-set that will make the radical freedom offered in such a world an attractive possibility. Many of us older people will find that same freedom chaotic and discomforting, if not downright disorienting, and it will be up to our children to teach us how to find our way in a world we were not born into.

All around us, the world is coming alive, infused with information and capability; this is the only reality for our children. It is what they learn every time they touch a Furby or build a Lego robot or play a video game, and it speaks louder than any lesson taught in any school, because the lesson is repeated—reinforced—with every button's touch. But it is up to us to rise to the challenge of a playful world, to finish the work of culture and change the nature of reality. It might seem, even after all of this, to be nothing more than a dream; but this is a book about dreams made real. So follow on, as we trace a path through a world that is rising to meet us.

CAVEAT FUTURE

No one can claim to know what the future will really be like. Periodically we review the predictions of the past for the present and find them woefully, sometimes hilariously inaccurate. We don't have personal helicopters—a dream of the 1930s—nor do we have computers we can converse with—a dream from the 1960s. It seems as though the best way to know what the future won't be like is to ask a futurist to predict it.

I don't claim to be a futurist, but I have been caught up in the midst of many of the developments I report upon in these pages. In a few cases, I happened to be in the right place at the right time. On other occasions, I've caught the scent of a new new thing and pursued it. That's in my nature: I'm a bit of a geek, with a fair helping of aesthete thrown in for balance. Over the last twenty years, the gyrations of my own career—from engineer to virtual reality entrepreneur to World Wide Web evangelist to chair of the Interactive Media Program at the University of Southern California's School of Cinema-Television—have reflected larger scale changes in our culture. I too am along for the ride.

Time can grant one the advantage of perspective; many of these new developments have been of interest to me for many years. I've had the opportunity to sit with these ideas and attempt to puzzle them out. What does the future hold? Rather than making something up—and probably getting it wrong—I'd rather show you what's going on right now, and then offer my own thoughts about where it might be leading. When you finish this book, you'll have the same facts in your possession and you can draw your own conclusions.

And, if you need more information, please come visit the companion website to this book, www.playfulworld.com. There you'll find most of my source materials, the documents and media I sifted through as I created this text. If you can spin these elements into a different story, I'd be glad to hear about it.

See you there.

I

INTELLIGENCE

1 FURBY FABLES

BATTERIES NOT INCLUDED (BUT A DICTIONARY IS)

"Me love *u-nye*."

A startling moment: when I get the four AA batteries into the casing, just as I tighten the screw, my little tiger-striped friend begins to chirp away. I almost drop it. It squirms in my palm, and I feel as though I've just done some veterinary surgery. Or perhaps some robot repair. Should I have used an anesthetic?

I consult the dictionary. *U-nye* means "you." "I love you," it's saying, perhaps grateful for the newly supplied power or delighted at the stimulation.

All of six inches tall, my new friend has a black mantle sweeping down its backside and a white belly that just cries out for a nice scratch.

"*Koh-koh*, please." He likes it and wants me to scratch him again. "Again," he says. And once more. Now he seizes the initiative and says, "*Kah a-tay.*" He's hungry—well, it *is* close to lunchtime—so I gently open his beak and depress his tongue. "Yum. *Koh-koh.*" I feed him again, per his request. And once more. Now he buzzes with the android equivalent of a gastric embarrassment, and says, "*Boo* like." He's a little fussy and doesn't want to be overfed. No like.

A moment later, he's happy enough, and sings a little tune. "Dee-dee-dee-dee-dee, dum-duh-duh-dum." Just in case he's said something important, I check the dictionary, but it's only a melody, a playful pause. So I pick him up, as I would with a small child, and send him flying through the air, carefully guided by my hands, doing cartwheels and somersaults and flying leaps. "Wheee!" he exclaims, and this needs no translation. My little friend expresses pure delight, giggling.

Back on the table, he sits while I make some notes. But he's a child in every sense of the word. Not content to let the world go by, he constantly interjects a bit of song or a few words; finally, after his repeated interruptions have been ignored, he pronounces the situation "boring."

My friend is really quite demanding. He wants to play, to enjoy my company, and he'll complain if he doesn't get all the attention that is his due. But I have words to write about him (though he doesn't know it), so he sits on the table, beside my computer, and continues to speak.

Finally I lean over his left ear, a broad black-and-white leaf that extends far past his body, and say, "Hello, my friend."

"Kah-dah boh-bay," he replies, startled. Translation: I'm scared. I guess I spoke a bit too loud. I pet him a bit, and he responds with a few sighs of pleasure. Then he goes back on the table. "Boring," he pronounces once again. He really *does* want 100 percent of me. Anything else and he'll act childishly petulant. This time he makes exaggerated snoring sounds. But I leave him alone. I have work to do.

Then my electronic friend sings the first bars of Brahms's Lullabye and goes to sleep.

THE HOTTEST TOY IN THE WORLD

The scene, repeated over and over, always went something like this. A harried parent, having made his or her way to an overcrowded shopping mall, fought through the stream of other, equally harried parents at Toys 'R' Us or Kay-Bee Toys or FAO Schwarz.

He'd pass stacks of Barbies and G.I. Joes and Tickle Me Elmos with a single goal in mind—a tiny furry talking toy. He could see the display, but as he approached, he found the shelves bare. In every store, on every website, the same sad tale.

In October of 1998, the Amazing Electronic Furby joined that rare class of toys which transcend mere popularity to become objects of fascination, cultural landmarks that tell us what our children are really drawn to. The Furby, from its introduction at the New York Toy Fair, caught the imagination of toy buyers, retailers, parents, and children across the United States, who collectively turned a little electromechanical gizmo into a bestseller.

Why? The rotund, wide-eyed furball is designed to be cute—sort of a cross between an overfed hamster and Gizmo from the film *Gremlins*. As a plush toy, it would have had a certain degree of success, popular with the under-ten set. But once four AA batteries have been put into its plastic bottom, the Furby comes *alive*—a yammering, demanding being who hungers for affection, food, and companionship.

Over a hundred years ago, Thomas Edison invented the first talking toys, adapting his mechanical phonograph to produce a pull-string doll that could recite a few prerecorded phrases. Over the twentieth century, toys grew more and more verbal, but still required some action on the part of a child, such as pulling a string, to produce a reaction. In 1980, calculator manufacturer Texas Instruments introduced the Speak and Spell, a high-technology wonder using microprocessor technology, together with voice synthesis, to teach children the alphabet and talk them through basic spelling exercises. With the Speak and Spell, the tables had been turned; now the toy took the active role; the child listened and reacted to it.

When the world of computers intersected with the world of toys, the concept of interactivity, of two-way communication between toy and child, ushered in a new universe of possibilities. Now toys could listen to children, observing patiently as they worked at various spelling exercises or games, and react to them personally, like a watchful parent, constantly assessing performance, gently extending the boundaries of the child's knowledge. This reactive intelligence produces something greater than the sum of its parts; the child often feels more engaged, and so works—or plays—harder.

This delicate dance between toy and child, now some twenty years old, has yielded only a few out-of-the-park success stories, most of these in educational toys, which drilled the child in various skills. Teddy Ruxpin, a mid-1980s toy dreamt up by Atari founder Nolan Bushnell, the father of pop interactivity, used sophisticated animatronic circuitry to bring a classic teddy bear to life. But Teddy Ruxpin hardly responded to the child, and while it could read a bedtime story to a toddler, it couldn't allow the child to participate in the storytelling. Though vastly more sophisticated (and expensive) than Edison's first talking dolls, it remained essentially a one-way device: Teddy spoke and the child listened.

The adventurous child can exhaust the limited possibilities presented by these so-called interactive toys in a few hours of play. Once the repertoire of activities has been explored, the toy ends up discarded, ignored. It holds nothing new for the active imagination, and the child moves on.

What if a toy could evolve, grow, and respond to the child, then demand that the child respond to it in turn? By the mid-1990s, such ideas, variants of which had been floating around the computer science community for several years, began to show up in a few different toys. The breakthrough product, however, was a watch-like device developed by Japan's Bandai Corporation, called Tamagotchi.

Billed as a virtual pet, the tiny device used a miniature LCD screen, similar to those found on most cheap wristwatches, to illustrate the life-state of the virtual pet contained within it. A Tamagotchi is born when it is first turned on by its owner, and thereafter it reacts much like a growing pet, requiring food, attention, and sleep. (Each of these virtual stimuli could be delivered through a few presses of a button on the face of the device.)

As the Tamagotchi grows up, it begins to mature, requiring less sleep, more food, and more playtime with its owner. If you forget to feed it, it will grow irritable and might even die. (When this happens, its iconic character on the LCD screen sprouts wings and flies off to heaven!) If you don't play with it, it will become ornery, just like any pet neglected by its owner. Fortunately, it will beep at you to indicate that it wants some food or just some attention. You can choose to ignore it, but that choice comes with a cost. *Here* was a concept that had never been embodied in a toy; unlike the animated playthings that had preceded it, the Tamagotchi could make demands and respond to neglect. Actions would have consequences.

Tamagotchi was a huge hit, first in Japan, and then worldwide, with the eight- to fourteen-year-olds, mostly girls who pinned them to their backpacks or belt loops or blouses. By engaging their desire to nurture, Tamagotchi gave these children an outlet they had been able to express only in the inert world of dolls. Barbie doesn't care if she's left alone for a month, but a Tamagotchi will likely expire if left alone that long. This simple fact had a profound psychological impact on these girls, who came to regard the toys as something nearly alive; less demanding than an infant human and certainly more portable than a puppy, but just as full of needs and programmed with the desire to engage them, an emotional manipulation more effective than any toy that had come before it.

A virtual pet, trapped inside a small plastic shell with a minimal display, could strike the fancy of an imaginative preteen, but wouldn't it be far better to create an entire, physical creature that could embody the principles of Tamagotchi? A real, physical toy could capture the hearts of younger children—along with older ones. Much of play is based in touch, in the feel of the real, and while Tamagotchi engaged the mind, it couldn't play with a child in any real sense.

Watching pet hamsters skitter across a bathroom floor, a designer named Caleb Chung began to conceive of an idea for a furry electronic pet, something that could take the simulation of Tamagotchi and make it real. After careers as a comic and a mime, in his late thirties Chung turned his prodigious talents toward toy design, and became a modern-age Geppetto, able to breathe the semblance of life into plastic and a few motors. Noting the explosive popularity of Tamagotchi, Chung reasoned that with a little work and some significant improvements, the virtual pet could become a real toy.

Roger Shiffman of Tiger Toys thought so, too. Over twenty years Shiffman and his partner Roger Rissman had built the Chicago-based toy manufacturer into a successful business, thriving even as their competitors went bankrupt or were acquired by toy giants like Hasbro and Mattel. They had scored a modest hit with Giga Pets, which resembled Tamagotchi in that a virtual pet expressed itself through its meanderings across a larger LCD screen. Like Tamagotchi, Giga Pets required the constant care and attention of their owners, displayed various needs, and responded to affection. However, Giga Pets were based on real-world animals, with names such as Digital Doggie and Compu Kitty, and behaved more like their namesakes; the on-screen dog could play fetch and the cat could wrestle with a ball of string. The toys presented the most salient features of each of their flesh-and-blood counterparts, but made them portable, manageable, and nearly as friendly. Still, locked inside their plastic casing, Giga Pets couldn't reach out to touch a child's heart in the same way that a pet or even a plush toy could. Software just doesn't feel real, especially to a small child.

So when Chung approached Shiffman with a prototype of the Furby, the stage was set for a revolution in interactivity, a virtual pet made flesh. Through a lightning-fast development cycle (most toys take at least two years to reach market, while Furby went from prototype to product in less than ten months) the team at Tiger Toys (and Tiger's new parent, Hasbro) found a sweet spot of interactivity and affect that charmed everyone—including its creators. Alan Hassenfeld, the president of Hasbro, loved Furby from the first moment he set eyes upon it, pronouncing it the coolest thing he'd seen in twenty-five years in the toy business. Media outlets from *Wired* to *Time* chimed in with their own rapturous opinions, and the vortex of hyperbole reached a frenzy on October 1, 1998, when Furby was officially unveiled at FAO Schwarz's flagship store in Manhattan.

Tiger Toys planned to manufacture a million Furbys for the Christmas season. Most of them had already been purchased, months before, by big retail outlets such as Wal-Mart and Toys 'R' Us. They sold out immediately everywhere they were placed on display. The $70 million advertising campaign that Tiger orchestrated seemed almost a waste of money. At $30, the toys were selling themselves. Parents couldn't refuse their child the hottest toy in the world.

That is, if they could find one. The shortage immediately spawned a black market in Furbys. Classified ads in the *Los Angeles Times* offered Furbys in several attractive colors for $200 apiece. In the online auction at eBay, bidding sometimes went as high as $400. Buyers spent a chill evening massed outside of a toy store in Santa Monica when it was reported that a shipment of five hundred Furbys would be released the next morning. Rumors of soon-to-be-released caches of Furbys flooded the Internet. Demand seemed insatiable. For every Furby unwrapped on Christmas morning 1998, four more could have been sold in its place.

Furby had struck a chord.

I'VE GOT FURBY UNDER MY SKIN

Furby's success represents more than just a fad. The Furby is the best example of a new class of toys—reactive, verbal, and engaging. While Tamagotchi and Giga Pets pioneered the virtual pet, Furby represents the first real outpost on the virtual frontier. To understand why, we need to lift the skirt of a Furby and find out what makes it tick.

Within a few weeks after Furby's release, hundreds of Furby fan websites sprung up. One of the most interesting of these is the Furby Autopsy site. Dedicated to the memory of Toh-Loo-Koh, a Furby who passed away under suspicious circumstances (which are never fully explained), the authors of this site decided that their recently deceased Furby should live on, in the interests of science, as an anatomical study of the animatronic toy.

Only a few stitches connect the Furby's fur to the plastic case that houses its electronic and mechanical components. Once these stitches have been cut, you can remove the fur by rolling it up to its head. Next, the fur covering its ears must be carefully released, and finally the entire furry surface can be pulled over the Furby's head. Now you have a skinned Furby, with its innards exposed.

If you're brave enough to complete this operation—without losing your lunch—you'll now see a small, almost cubic block, densely packed with gears and electronics. This is the real Furby, a robot with Ping-Pong ball eyes and a circuit board for a diaphragm. The authors of Furby Autopsy describe the gear and cam system, which activate all of Furby's moving components, as a work of genius, simplicity itself. A single driveshaft controls all of Furby's motions, unlike Disney's animatronic monsters, which have thousands of independently moving parts. Various rotations of the driveshaft control different activities, such as blinking the eyes, dancing, or wiggling its ears.

The motor turning the driveshaft is mounted on a printed circuit board, similar to those found in other consumer electronics, such as a Walkman, and is itself controlled by a host of electronic components also located on the board. This is the real source of Furby magic, where its complex programming produces a simulation of life.

All interactive toys must have at least two different types of components: sensors which allow the toy to know what is happening in the environment around it, and affectors, which allow it to respond to the environment. There are sensors to mimic every sense that we possess: light sensors, microphones, switches that turn on or off when tilted (a sense we have in our inner ears), pressure switches (which sense touch, much like our skin), and so on.

The Furby doesn't completely duplicate the innate human senses (for example, it can't taste or smell), but it can sense light and dark: a third eye located above its peepers detects light sources. It can hear sounds through a microphone located within its left ear, and it can be tickled or petted by means of pressure sensors located on its stomach and its backside. Finally, a tilt sensor located behind its extra eye lets it know if it's upright, being tipped upside down (which is how you wake a Furby after it's gone to sleep), or just being somersaulted through the skies.

All of these sensors are wired into a set of microprocessors contained on the circuit board. As the signals from each flow in, they're sampled for their current values and then fed into a complex computer program that brings the Furby to life. While the mechanical design of the Furby may be genius, the Furby's program is the real work of art, carefully crafted to make the creature seem as lifelike as possible, given its limited capabilities.

A conservative estimate of the Furby's brainpower puts it at one ten-billionth of our own human capacity. Why, then, does it seem so alive? Because it plays upon human psychology, our desire to anthropomorphize—to see the inanimate and nonhuman as human. We do this all the time: look at how people behave toward their beloved cars or boats or computers, treating them as people, with feelings that can be hurt. Yet these devices don't respond to that affection. The Furby does, using two techniques—facial expressions and its verbal ability—to create the illusion of real life.

Over the last few years, researchers working in the field of cognitive science have begun to discover just how important facial expressions are to human beings. It seems as though we're able, from the moment of birth, to recognize facial expressions on those around us. This instinct probably helps mothers to bond to their children and helps children recognize their caregivers. It also allows mothers to recognize and respond to their children's physical and emotional states long before the children can express these verbally. People often note a mother's uncanny ability to read the mind of her infant; in part, she's interpreting facial expressions.

The Furby can produce faces that look a lot like wide-eyed wonder, anger, sleepiness, and playful joy. We don't need to learn what these expressions mean; they're reflections of our own innate understanding. When we encounter them, we immediately believe that these expressions imply a human-like depth to the being or toy displaying them. It's all in our own minds, just an illusion of humanity. But that doesn't seem to matter. Our emotions carry us away, and it becomes easier to believe that Furby is alive.

With the probable exception of whales and dolphins, human beings are the only species blessed with a verbal consciousness, the ability to translate the details of our thoughts into oral communication. Many animals communicate, to find mates or warn others against predators, but no other species has been shown to communicate so broadly (and about such trivia) as human beings. It's one of the things that makes us human. To be human is to live in near-constant communication.

Of all of the Furby's affectors, the most indispensable may be its voice box. Although it simply plays prerecorded sounds, the Furby has an ability with language unlike any toy that came before it. Furby has a built-in vocabulary of a few hundred words, many of them in a homegrown language known as Furbish. But rather than just uttering single words or canned phrases, the Furby has programming that allows it to combine these words into a pidgin—a simplified, bare-bones language similar to others that evolved in the multilingual trading ports of the South Pacific in the seventeenth century, where Portuguese and English mixed with a hundred native tongues. This stripped-down syntax allows a Furby to utter at least a thousand different phrases of all types.

While not exactly encyclopedic, a Furby's repertoire of sayings fairly replicates the linguistic abilities of a child in its first years of speech. Its ability to voice its needs and desires gives Furby an implied depth. The subconscious mental calculation goes something like this: only humans speak; only humans can take language and twist it into new forms; therefore anything that can use language like a human must be (nearly) human. When Furby's voice is combined with its range of facial expressions, it becomes easy to see why people take to it instantly: it seems real enough to be human, to be accorded the same depth we grant other people. It seems a worthy object of affection. Children give their affections broadly, but adults tend to be much more circumspect; and yet Furbys capture the hearts of adults as readily as they do children.

What do Furbys talk about? This is perhaps the most interesting and lifelike aspect of the toy, because Furbys have *needs*. Like the innards of Tamagotchis and Giga Pets, Furby's software has been crafted with a sense of purpose, of requirements that must be met in order to keep the Furby healthy and happy. For example, a Furby must be fed on a regular basis—not real food, but rather, by opening its yellow plastic beak and depressing its tongue. A Furby almost always responds with a "Yum" when fed, and though it might occasionally break wind if it's been overfed, it needs to be fed daily or it will catch a cold. (It's true: the Furby will lapse into near-continual sneezes if it's been consistently underfed.)

Once its appetite has been sated, the Furby wants what almost any pet wants—the attention of its owner. It might ask to be tickled or to be taken to bed and told a story. Because of its sensors, it knows when it is being stroked or read to and can respond in kind. (Furbys don't understand English, though, and can't respond to the details of a story. That's a possibility reserved for science fiction, and perhaps some toys of the future.) The Furby has a soul whose essence is play; its greatest delight, as judged by the sounds it makes and the expressions upon its face, comes when it is deeply involved with and responding to a child.

Furbys *evolve*. Though this may seem a ridiculous claim to make for a mechanical device, an important part of the Furby's programming concerns its evolution, and a memory of its own interactions with its owner. When a Furby is first brought to life, it speaks a patois comprised mostly of Furbish, with only a few English words. But as the Furby grows older—that is, after it has been played with over a period of time—more and more English words work their way into its vocabulary—just as you might find happening in a child during the transition from baby talk to intelligible verbal communication. Although this evolution was an easy thing for Furby's programmers to create, it adds a certain depth to the experience of owning the toy. Like a real pet, the Furby remembers its interactions with its owner, particularly those activities that bring delight to its owner. (This is easy to determine: if a child likes a certain activity, he or she will repeat it. Endlessly.) As it grows up, the Furby is more prone to engage its owner in the behaviors that bring the two together joyously. And since each owner has a unique set of likes and dislikes, no two Furbys will evolve along the same lines. By the time it's fully mature, each Furby is a reflection of its owner, a combination of mechanisms, programming, and evolving behaviors that together produce a compelling simulation of real life.

Finally, *Furbys can learn from each other.* That mysterious third eye located above its orbs contains a light sensor, but it also contains an infrared transceiver similar to what you might find in a television remote control and the television it's controlling. When Furbys are placed in proximity with each other, they can communicate silently, using invisible pulses of light. This is the equivalent of a Furby local area network, and all Furbys within the line of sight will likely begin a group conversation, asking each other to dance or playing a learning game where each teaches its peers some phrases of a song. This only adds to the perception that there's something under the skin, a consciousness very much like our own.

Of course, it's all an illusion; the Furby is not conscious, at least not in the way we think of ourselves as being conscious. But the Furby provides enough of a vehicle on which we can project our own ideas of what constitutes a human being that it's very difficult not to treat them as real entities, with feelings, needs, and desires.

Is there any doubt about why they've suddenly become so popular?

MAN'S BEST FRIEND

After Christmas came and went, and a million lucky children had unwrapped their Furbys, stories about the little toy began to work their way into the media. It wasn't enough that the Furby had become the object of desire for so many holiday shoppers; the world's fascination lived on. As people grew more attached to their Furbys, they told or read stories about them, and as each tale reached another set of ears, a whole new set of myths grew up. A few of these stories made it into the media; they paint a picture of how the Furby captured the hearts and minds of children across America.

In early January, as people went back to work, CNN reported that the National Security Agency had banned the playful pal from their Maryland headquarters as a security risk. It seems that the NSA believed that the Furby had the ability to record sound—just a second or two—from the microphone embedded in its ear. This, reasoned NSA officials, made it possible for Furby bearers to smuggle state secrets off the premises. Recording devices are banned in the halls of their super-secret Fort Mead center, and employees are searched as they enter and leave the facility. When Furbys began to show up on employee desks, they were ignored as just a toy—for a while.

Before long someone at NSA security learned of the Furby's microphone, and from this surmised (incorrectly) that the device could record human speech. Imagine spies feeding their secrets into the seemingly harmless pets. Furby could be a fifth column that might jeopardize American security! So a directive came down from NSA officials stating that "personally owned photographic, video, and audio recording equipment are prohibited items. This includes toys, such as 'Furbys,' with built-in recorders that repeat the audio with synthesized sound to mimic the original signal. We are prohibited from introducing these items into NSA spaces."

Tiger Toys immediately issued a statement declaring that the Furby, while it could "hear," couldn't record sound. But the idea of a traitorous Furby had caught on, and soon after the NSA ban, NASA's security office announced a similar prohibition. It all seemed like a case of the bureaucratic willies, brought on by an enormously engaging toy of unknown capabilities, the perfect background onto which people could project their own fantasies, pleasant or paranoid.

NSA employees, upset at being deprived of playtime with their animatronic pets, began to complain on the agency's electronic mail and bulletin board systems. "Damn! Next they'll tell me my rubber ducky has to go," quipped one. Others wondered what all the fuss was about, saying, "I don't think we have a problem with nonemployees wandering into NSA with a Furby 'under their coat.' " As the controversy grew, embarrassed NSA officials issued a final statement, "Please cease and desist posting on this topic immediately," thus summarily ending all discussion.

To this day, no one has been able to demonstrate that a Furby can record sound, but the NSA ruling stands. Just in case.

A few weeks later, another story appeared on CNN's website, this one scooped from the pages of the *National Enquirer*, that bastion of fair and accurate news reporting. A tale that would normally have been ignored by the mainstream press—bizarre as any alien abduction or three-headed baby—was picked up and broadcast *because* it involved the Furby.

Here's the report in full from the *Enquirer*.

Fire! Furby Saves Soap Star's Life

Former "Young and Restless" actress Candice Daly and her boyfriend were saved from a fiery death by her Furby doll!

"We call him our little angel now," declared comic Quentin Gutierrez, boyfriend of Candice, who played Veronica Landers. "That Furby saved our lives."

Quentin sprang out of bed when the Furby, a Valentine's Day present for Candice, started squawking at 3 a.m., and he found a chest of drawers in flames!

"Earlier Candice and I had dinner, drank some wine and fell asleep. We forgot all about candles we had left burning," the 37-year-old comedian told The ENQUIRER.

Candice loves that Furby and loves to sleep with it, and thank God she does! The Furby goes to sleep when it's dark and wakes up and starts to talk when the lights come on. It makes noise like a rooster crowing.

"The light of the flames woke up the doll and it started talking.

"It kept yelling 'Cockadoodledoo' and talking baby talk, and I

> finally woke up enough to look around, and I saw the flames! A candle had burned down and set fire to a chest of drawers. We were only seconds away from a devastating fire which would have killed us both!
>
> "My two fire alarms didn't make a sound, but Furby was a real life-saving hero! I grabbed some underwear and smothered the fire.
>
> "He's the greatest toy on earth!"

Every quality attributed to the Furby by the *Enquirer* article is in fact true. (Yes, folks, some of the facts reported in the *Enquirer* aren't made up. I'm as surprised by this as you are.) The Furby will go to sleep when the lights go down (remember, it has a light sensor in its third eye), and in the morning, when light returns, it will wake up with a Furbish "Cock-a-doodle-doo!" So, in fact, it is possible that this incident did occur, that Candice Daly's Furby did save her life, and the life of her quick-to-react boyfriend Gutierrez. Which is probably a first, both in the brief history of the Furby, but also in the much longer history of toys.

It rates barely a yawn when we hear a stirring story about how a faithful dog saves its master's life from a fire, intruders, or some other disaster. Man's best friend has a long tradition of taking care of us—even as we have taken care of them. It's the bargain we worked out many thousands of years ago, when we first domesticated these cousins of wolves. Within a few weeks of its arrival, people were willing to cut Furby in on the deal, at least theoretically.

If a creature could respond to you, could talk to you in your language (or, if you cared to learn it, its own), and generally act as best friends for life from the moment you popped in its batteries, perhaps it could rival the flesh-and-blood pets for the role of human companion. An intriguing idea which, though left unsaid at the time, must have been floating around in the public mind as this report reached the airwaves. And if this report had talked about any other toy besides the Furby, it would have been dismissed as ridiculous. (It might have made it into the *Enquirer*, but CNN? I think not.) But in the brief span between October 1998 and February 1999 so many of our own human qualities had been applied to the Furby—more anthropomorphizing—that they seemed nearly human, and certainly far more intelligible than dogs or cats.

The nation had a new pet—and a new protector.

The last of these stories is by far the most serious, and leads us into a deeper discussion of the meaning and future of the Furby, because it directly addresses the essential human relationship the Furby plays upon to produce its intoxicating effects. On July 6, 1999, Wired News, which had earlier hailed the Furby as a revolutionary toy, reported that a woman from Blacksburg, Virginia, claimed her son's Furby had taught him at least half a dozen new words.

This wouldn't be an outstanding claim in any normal circumstance; toddlers pick up words wherever they are uttered, but Lisa Cantara's son, C.J., is mildly autistic. Although much remains a mystery about autism, it is known that autistics have trouble building relationships with people (or pets), that they acquire language very slowly, and that they thrive on repetitive behaviors. (Where a normal child might tire of some behavior after the thirtieth or fortieth go-round, the severely autistic might still be deeply interested well into the thirty- or forty-thousandth.)

For an autistic child, the Furby might be an ideal pet; it speaks to a child in English, might even inspire that child to learn Furbish, and enjoys constant, repetitive stimulation, never growing bored with the attention. Cantara was quoted as saying, "It increased his vocabulary and helped his speech. He started saying 'I'm hungry,' instead of just 'Hungry.' The Furby taught him that."

She noted her son's rapid improvement in language skills after he was given the toy. "He became very attached to it. He carried it around with him everywhere and started to mimic it. He talked to it a lot. He treated it like it was a real creature." Despite his difficulty connecting with other human beings, C.J. was able to establish a bond with the animatronic pet. Through repeated interactions, the Furby was able to improve C.J.'s communication abilities with others. "I only wish I could get Furby to say more words," Cantara wrote to an online bulletin board of Furby hackers—people who, like the creators of Furby Autopsy, take their Furbys apart to better understand what makes them tick. Cantara pleaded with them for help. "It's been a great learning tool for my son, as odd as it might seem."

In fact, by midsummer 1999, similar reports were streaming into Tiger Toys' Chicago headquarters. The odd had become commonplace. In the same article, Tiger Toys spokeswoman Lana Simon commented, "It's a really magical toy. We've heard from the parents of handicapped children and nursing homes, where they've been given to Alzheimer's patients. I've heard of them causing reactions in people that otherwise show no movement. It registers with them, it stimulates them."

The magical Furby, which mimics human behaviors through facial expressions and vocalizations and which portrays at least a fair simulation of consciousness, seems to be able to connect with people who would otherwise be disconnected, cut off from the human world, because of infirmity or circumstance. It gives people the companionship of a pet without all of the allied responsibilities, and it provides an English-speaking counterpoint to their quiet lives. People who couldn't manage a dog or cat can easily play with a Furby, to the delight of both. Furby feigns the same interest and delight and makes some of the same demands (though with a much lighter burden). Just as pets have been proven to help people maintain their contact to the world at large, the Furby works the magic of relationship to create a healthier, happier person.

We know that children need toys to stimulate their imaginations, to help them along the flights of fantasy that will lead them to determine their likes and dislikes, their social skills, and even their choice of careers. Adults often consider themselves above toys; in fact, a line in the Bible speaks directly to this: "When I was a child, I talked like a child, I thought like a child, I reasoned like a child. But when I became an adult, I set aside childish ways" (1 Corinthians 13:11). It's almost an embarrassment for an adult to admit a fondness for a children's toy—beyond a nostalgia for toys of the past, that is. But Furby seems to have transcended the category of toys for children, and as a faithful and loyal companion, might be on its way toward a permanent place in our culture. The way it engages us, the needs it seems to satisfy, tell us something about ourselves and about the future of our relations with the artificial world. Furby is a starting point, a launching pad into a new tale of the long story of man and machine.

In order to understand where Furby is going, what it might look like by the time our millennial child has reached her sixth year, we need to rewind the tape of history just a bit, and understand how the development of artificial intelligence will inform the future of the Furby. From a study of a toy we'll transition to a story of big science, to see how each converges upon the other.

2 GROWING INTELLIGENCE

THINKING MACHINES

In the middle of the eighteenth century, Austrian empress Maria Theresa received an unusual gift: a mechanical contraption that could play a game of chess. Nicknamed "The Turk" because of the life-size turbaned automaton that moved its pieces across the chessboard, it quickly became the sensation of European salons. Wolfgang von Kempelen, the inventor of the device, would wind up the apparatus, and, with a great deal of mechanical noise, it would proceed to challenge any opponent. The Turk usually won, besting such luminaries as King George III of England and Benjamin Franklin, both of whom possessed formidable minds for the game.

At the beginning of the modern era, people were very suspicious of such animate machinery, and von Kempelen offered no details about the construction of the machine, preserving his secret and his notoriety. But its wooden case, about five feet in length and another three in depth and width, seemed large enough to contain a human being, so von Kempelen, during exhibitions, would open it up with a dramatic flourish, to reveal only a mechanism, nothing more.

A machine that thinks? This was a wonder of the Enlightenment! Too bad it wasn't true. Although Europeans might have been wise to trickery, von Kempelen knew the magical arts better than his patrons. In an exposé entitled *On the Chessplayer of Mr. Von Kempelen and Its Forgery*, published twenty years after the empress took possession of her prize, Joseph Friedrich revealed that a human player was indeed hidden in the works, a *midget* grand master of chess crowded into a false panel *beneath* the mechanical components von Kempelen delighted to reveal. The illusion worked perfectly because, once von Kempelen had "proven" that no human being controlled the apparatus, skeptics were ready to believe the inconceivable.

Despite his fraud, von Kempelen excited the scientific minds of his day—Franklin being only one example—and The Turk could be considered as the igniting spark in the field of artificial intelligence. The idea that human-like intelligence could exist outside of human beings had previously been confined to the less-than-scientific regions of folktales, of grinding mills that could magically produce any desired substance or magic carpets responsive to verbal commands. The idea of charmed objects has a long history in world mythology, but in the Age of Reason these stories became little more than tales for children, bedtime stories.

The Industrial Revolution revitalized the idea of the thinking machine. In France, Joseph-Marie Jacquard designed looms that read a series of punched cards to produce a complex pattern of weaves. James Watt perfected the steam engine with the invention of the governor, which allowed the output of the engine to be fed back in as its input, thereby keeping it from overheating and exploding. George Boole proved that all mathematical operations could be reduced to a near-endless string of zeros and ones—a formal theory to complement Jacquard's cards—while Charles Babbage tried to create a Difference Engine that could perform any mathematical operation, however complex, by reducing it to a set of instructions that could be programmed by twists on a dizzying array of levers and valves. By the end of the century, adding machines found a place in the business office, using mechanical rotors and gears to calculate the ever-increasing flows of physical materials and money that were the direct consequence of the Machine Age.

With the discovery of the electron by J. J. Thomson in 1897, the second phase of the Industrial Revolution—the Age of Electronics—began in earnest. In a few years, vacuum tubes, which harnessed these minute particles of energy, entered common use, finding their way into radios and all manner of electronic control devices. Now scientists contemplated the design of circuits, complex arrangements of electronic devices that can produce a wide array of different effects, such as amplifying sound or transmitting and receiving radio waves.

Perhaps the most far-out application of early electronic research concerned signal processing, the analysis of how to recover information from an input that had gotten "noisy," cluttering up the precision of its values with a messy stream of random data. This began as an exercise in mathematics, but, with the advent of the Second World War, it took on incredible importance. These same circuits that had consumed scientists' fancies were the key elements of a new system, known as radar, which would allow for the detection of airplanes many miles away, in absolute darkness. The radar system sent out a stream of pulsed radio waves, then read their reflections, filtering its responses from a jungle of noise. It pushed the mathematics of electronics to their limits and gave birth to a new field of scientific research.

Among the outstanding scientists of the twentieth century, Norbert Wiener stands with the greats, shoulder to shoulder with Einstein, Salk, or Curie. He's gotten relatively little press over the years, and most people are unaware of his contribution to the world. But the prefix *cyber-*, now in continuous use, is his gift to us. One of the mathematical geniuses behind radar, after the war Wiener published a short book entitled *Cybernetics: Command and Control in Man and Machine*. Going back to classical Greek (which Wiener could read and write before age ten), he found the word *kybernētēs*—which translates as "governor"—and applied it to all systems that control (that is, govern) themselves. Watt's steam engine used a governor to regulate its output, and Wiener's governors, which he called feedback, used electronic components to produce circuitry that could regulate itself, cancel noise, and produce intelligible information from very faint signals.

Wiener understood the full implication of his work, and *Cybernetics* details his theories about human biology, the regulatory mechanisms that keep us alive—theories that have in large part been proven correct—and outlined the architecture of sophisticated machines that might someday be able to produce the appearance of intelligence. With enough inputs and enough feedback, Wiener surmised that a machine could behave intelligently, could modify its own behavior to suit present circumstances, and could respond to nearly any possibility. This range of flexibility had never been the domain of the machine, which, at the time of Wiener's writing, remained a fixed, single purpose device.

Within a small community of scientists, social scientists, and anthropologists (Gregory Bateson and Margaret Mead were both great admirers of Wiener), *Cybernetics* prompted a conceptual revolution, giving birth to fields as diverse as systems analysis and immunology and modern information theory. But Wiener's ideas were dense, compressed; it was said of his graduate lectures at MIT that he was the only one in the room who understood them. For nearly fifty years his radical ideas about intelligence in the machine remained influential, but mostly unrealized.

On the other side of the Atlantic, another scientist worked along lines similar to Wiener's. Oxford mathematician Alan Turing, assigned to Bletchley Park, the super-secret installation charged with breaking Nazi encryption, invented a predecessor of the modern computer to automate the highly repetitive task of testing all possible ciphers for a given message. This device probably won the war for the Allies. With the codes broken, it was possible to locate U-boats on the seafloor, bomb supply ships with deadly accuracy, and even follow the confused flurry of German messages during the first hours of the D day campaign.

When not consumed with breaking the code, Turing's mind continually churned on the subject of intelligence in both machine and man. In the 1930s, he wrote a short paper that came to define the possibilities for computing. His Turing machine is the conceptual precursor for all modern computers. In Turing's perfect mathematical world, computers could solve any problem put to them, given enough time and resources. Could this, he wondered, create the semblance of intelligence? He then asked himself a more fundamental question: what *is* intelligence?

That's a substantial question, so Turing knocked it back down to manageable dimensions. If we can't know what constitutes intelligence, can we at least create a test for it? From this, the Turing test was born.

Imagine yourself online, in one of the thousands of chat rooms or instant-messaging systems that now clutter the Internet. At first you're by yourself, but you soon find yourself joined by another person, who introduces himself (or herself) to you. The two of you engage in conversation. You talk about anything and everything: the weather, your favorite sports team, maybe even the stock market. Exactly when, in this conversation, do you decide that this person is a human being? You probably assume that anyone in a chat room is another human being (aliens aside), so from the beginning you might be inclined to believe that this other entity is a real person. But what if his responses become stilted? What if you begin to sense a lack of depth? What if you feel that you're not talking to a person at all, but something that's doing its best to act like a person? How long would it take for you to make that determination, if you could ask any question you wanted?

This, in essence, is the Turing test. We might not know what intelligence is, said Turing, but we know it when we see it. That fact alone is enough to give us a sure tool to make a decision. After all, if you can't tell the difference between a human in a chat room and a machine in that same chat room, who's to say that the machine isn't just as intelligent as the human?

The Turing test is the gold standard for *artificial* intelligence because it's based on an assessment of human intelligence. In the understanding of intelligence, we can only look to ourselves. And this is the biggest paradox artificial intelligence has ever faced.

OPEN THE POD BAY DOORS, HAL. PLEASE?

In 1956, computer science researcher John McCarthy invited scientists from around the world to participate in a summer research project on artificial intelligence (or AI). McCarthy coined the term for the conference, and it stuck. Although the conference was not a great success in technical terms, it did bring the principal minds of the emerging field together and laid the foundation for future research. So much was unknown about constructing an artificial intelligence; then it was thought that natural (human) intelligence should serve as the starting point for their investigations.

Very little was known about natural intelligence. We think, we know we think, we can even break some of our thoughts into chains of reason. How these chains of reason emerge from our consciousness and how the events of perception (a flashing light, for example) come to mean something (such as an alarm) was still unexplored territory.

Rather than hypothesize endlessly on the processes of consciousness, researchers set to work building tools to test theorems they would develop over the course of their work. The first milestone—again, by pioneer John McCarthy—was the invention of one of the first computer languages, LISP, which because of its elegant structure could be used to design programs that could modify themselves. Wiener's concept of feedback, of systems regulating themselves—a concept essential for these early researchers—had been captured in a piece of software that would serve as the bedrock of AI research for the next forty years.

With a computer language capable of expressing their models of intelligence, the experiments began. And government grants rained down, driven by military leaders who dreamed of automating much of the decision making associated with the Cold War, believing an intelligent computer far more vigilant than even the most able officer. The principal benefactor of this largesse was the Artificial Intelligence Laboratory at the Massachusetts Institute of Technology in Cambridge, Massachusetts, which became the focal point for AI research, its leading light, home to such luminaries as Marvin Minsky (who will appear repeatedly throughout this book), Seymour Papert, and Richard Stallman.

Early on, AI made enormous strides. Working under Minsky, a graduate student named Terry Winograd spearheaded a project called SHRDLU, which showed that a computer could be taught to understand the relationships between a set of geometric figures arranged in a small space. With SHRDLU, the computer could respond intelligently to simple English commands, such as "Put the blue cone into the green box." Another program could solve word problems involving algebra. Yet another could interpret simple English sentences.

An article from *Scientific American*, December 1949:

> **Universal Translator?**
>
> If machines can be built to count, calculate, play chess, even "think," why not a machine to translate one language into another? British workers are planning a translator based on the storage or "memory" apparatus in a mathematical machine. After "reading" the material to be translated by means of a photoelectric scanning device, the machine would look up the words in its built-in dictionary in the instrument's memory unit, and pass the translation on to electric typewriters.

At the inception of research into artificial intelligence, many researchers believed that language translation—the ability to convert English into Russian, for example, and vice versa—would yield quick and plentiful results. But computer-based translations took well-formed sentences in one language and converted them into doggerel in another. It was often ridiculous. For example, "The spirit is willing, but the flesh is weak," twisted itself into "The vodka is good, but the steak is lousy." The difficulty lay in *context*. Human languages, laden with ambiguity, with subtle shades of meaning and variations of tone, befuddle even their native speakers. A computer, looking just at the words themselves, ignorant of their broad definitions and common usages, couldn't make sense of them once they'd been strung into complex sentences. Russian, a language well known for its subtlety and internal play, just couldn't be programmed into a computer; too much of the sense of Russian language derived from the history and culture of the Russian people.

Games fascinated AI researchers, and many hours went into programming rule-based games such as checkers and chess. An engineer at IBM named Arthur Samuel built a program that played checkers, learned from its mistakes, and soon surpassed the abilities of its teacher. Chess, a much more complex set of rules and interdependencies, couldn't be learned, at least not in the same way. Too many possibilities were open to the chess player. Too many exceptions came into play, and too much specific knowledge was necessary to create a useful model for the game. In the end, AI researchers adopted a brute-force approach to chess: the computer examines every possible move, in every possible combination, and selects one that seems to point toward a victorious outcome. It's not intuition, just an educated guess, and not a particularly intelligent strategy. (Given that one of the world's fastest computers has beaten the reigning grand master only once, it's clear that even this approach, which could never be mimicked by a human being, has its shortcomings.)

Optimistically, AI researchers ignored the obvious limitations in their own work and focused on the future. Every researcher polled in the 1960s stated that intelligent machines were less than twenty years away. They said the same thing in the 1970s and in the 1980s. But the target seemed to grow no closer; even as the simplest problems yielded to research, a dozen intractable problems appeared on the horizon. And there were some deceptive victories.

In 1964, Joseph Weizenbaum, a professor at MIT, wrote a simple LISP program which he named Eliza. The program did very little but examine English sentences for their subjects, objects, and verbs. That isn't particularly difficult, even if the context surrounding each of these is difficult to decipher. Weizenbaum did something very tricky: he gave the program a back story. He christened Eliza an electronic therapist (of the Rogerian school—they listen to your words and feed them back to you) and had Eliza regurgitate its input as output. Many versions of Eliza live on the World Wide Web, and here's a real session, from one of the Eliza online websites:

> I type, "My computer has broken." (As it did, while typing this chapter.)
> Eliza responds, "DO COMPUTERS WORRY YOU?"
> "No. I write books about the uses of computer technology."
> "DON'T YOU THINK COMPUTERS CAN HELP PEOPLE?"
> "In certain situations, yes."
> "I SEE."
> "What do you see?"
> "WHAT ANSWER WOULD PLEASE YOU MOST?"
> "An honest one."
> "PLEASE GO ON."
> "I'd like to know why people think you're a real person."
> "I'M NOT SURE I UNDERSTAND YOU FULLY."
> "When you were created, people mistook you for a human therapist."

"TELL ME MORE ABOUT THAT."

"Weizenbaum let students play with the program, but it turns out that they took the therapy seriously—claimed that you helped them with their personal problems."

"WE WERE DISCUSSING YOU, NOT ME."

"But I am trying to understand your allure."

"ARE YOU WORRIED ABOUT SOMEONE ELSE'S ALLURE?"

"Yours, yes. Because you're not a real person. Not even really intelligent."

"I SEE."

"Do you? Really?"

"WE WERE DISCUSSING YOU, NOT ME."

And so on. Eliza doesn't seem very intelligent and certainly wouldn't pass a Turing test, but to Weizenbaum's surprise and horror, people anthropomorphized Eliza, imagining it to be some distant therapist, typing away as it tirelessly listened to their complaints. Even when users knew it was just a computer program, the magic held. Eliza, like the Furby, provided enough of the right hooks to engage its users, using clever tricks of language to make itself seem intelligent. Almost human.

By the mid-1970s, artificial intelligence seemed poised to be the growth industry of the high-technology sector. Companies such as IBM and Digital Equipment Corporation poured millions of dollars into the development of systems designed to mimic human reasoning. But the end was near. DARPA, the Defense Advanced Research Projects Agency, which a few years before had funded the creation of the Internet, sponsored a program to develop a smart truck, an artificial intelligence robot that could perform battlefield operations. (Not a bad idea, mind you—let the machines do the fighting.) But the task was too complex, too context dependent, and ultimately unprogrammable. All of AI's successes hadn't cracked the biggest nut of all: how to adapt to a rapidly changing situation. In 1989 DARPA canceled the project and delivered a deathblow to an already contracting AI industry.

Anyone who has ever used a voice recognition system already knows why the smart truck failed. Voice recognition, begun at Bell Labs in the 1950s, was supposed to supplant keyboards as the favored mode of interaction with computers. Marvin Minsky brought it to the popular imagination when he suggested the idea to Stanley Kubrick as the filmmaker was working on *2001: A Space Odyssey*. Hal 9000 embodied the best ideas gleaned from the community of artificial intelligence researchers. By the millennium, they reasonably supposed, we would all be talking to our computers, engaging them in human conversation.

This proved to be as difficult as computer language translation. Computers don't grasp the multiplicity of ways we use words to express what we mean. A statement like "Outside, she saw the outside of the saw" could drive a computer into digital apoplexy. Even computerized dictation is frequently inaccurate, and it doesn't even require the computer to understand anything beyond the sounds of the words. Despite the exaggerated claims of computer executives like Bill Gates, who claim voice recognition is "just a few years away," don't believe the hype. We've been hearing that since the 1960s.

All of these failures relate directly to one point, which brings us back to an image from the first days of AI—there's nothing inside a computer to assign meaning. We want to *believe* that a Furby is alive or that Eliza really understands our problems. In each case we think we've encountered a device that has somehow acquired the ability to perceive the meaning of the world. But it's only a façade, just clever programming. We have only the vaguest idea of how we do it ourselves. For a computer to serve as language translator or even a reasonable dictation device, it would need a little person inside the box, carefully removing the ambiguities from our speech, from our thoughts, from our actions. Brute-force approaches won't work; a certain degree of subtlety—of *intuition*—is required. And intuition remains stubbornly beyond the domain of reason, integral to everything we say or do, but unfathomable to a computer.

AI died in the 1980s because researchers underestimated the truly unique gift of human intelligence. The thinking machines of science fiction remain just that—fiction. The final proof of this was the Gulf War. The smart bombs, blessed with a kind of artificial intelligence, required human pilots to direct them to their targets. The automated warfare so eagerly sought by the military became war at a distance, a telewar, with missiles the robot puppets of human controllers. Instead of demonstrating the great leaps in artificial intelligence, the war only reemphasized the importance of the human in the machine equation.

But just as the corpse of AI was laid to rest, other intelligences began to grow.

INSECT ZEN

Allen was simple. Some even called him stupid. If you left Allen alone, he might sit in the center of the room, doing nothing at all. Then, as if a sudden thought had possessed him, he might roll into the corner and sit there. And if you went over to Allen, he'd show himself as a bit skittish. If you got too close, he'd run off and find another corner to settle into.

Allen wasn't a looker, either. In fact, he rather resembled a bar stool on wheels, with a lot of electronic guts exposed in his midsection. But despite being both ugly *and* stupid, Allen performed a task no other robot had ever been able to do—he could avoid you. This might not seem like much of an accomplishment, but in the history of artificial intelligence, it was something of a watershed.

In the 1970s, before the death of AI, millions of dollars had been spent to develop robots sophisticated enough to avoid obstacles in their path. One of these, nicknamed Shakey, could be given commands via a computer, such as "Go over to the cube." After some serious thinking, Shakey would maneuver through a field of objects, find the cube, and move toward it. This was reasonably impressive, to be sure, but Shakey worked only in a specific environment, with large objects painted primary colors, a floor of a certain light color, walls of a darker color, and a black baseboard. It all had to be very well lit. You see, Shakey had an eye, which could take in the scene, recognize each of the objects in it, build a "map" of these objects, recognize and select its destination (the cube, in our example), then calculate a path to the object. Once all of that had been "thought" about, Shakey would roll its body along the path and toward its goal.

Shakey followed a particular cycle—observe, map, plan, and move—which, it was thought, resembled the way a human would approach the same problem. AI researchers imagined that a human being goes through this entire process, albeit at lightning speed, every time he moves about a room. But a human being can deal with changes, such as the movement of furniture, which would completely confuse Shakey. If any of the objects were tilted even slightly or if the lighting wasn't perfect or if one object blocked another from view, Shakey would fail to reach its goal. Unlike a human being, Shakey couldn't adapt or generalize its behavior.

Other living creatures, all the way down to insects, do this all the time. Consider a housefly, which manages to weave its way through the air to land upside down on a door frame or at an eighty-degree angle on a lampshade or on a moving object. That fly, with a very simple nervous system, can perform tasks like this nearly perfectly. But a multimillion-dollar robot couldn't even come close.

This comparison obsessed Rodney Brooks. In 1983, as a junior faculty member at Stanford University, Brooks was invited to give a talk about how computers could use vision to learn about their environments. On the appointed day, the tall, brash Englishman opened his talk with a searing indictment of AI's unquestioned truths. Everything you know is wrong, Brooks claimed. The proof is in the fact that it doesn't work. Expressing his frustration with AI research, he suggested that instead of building computers that knew everything about their environments, he'd rather design computers capable of learning from their interactions with the environment. Brooks didn't know if he could build this kind of learning computer, but he wanted to try.

This observation, born out of a sense of futility, brought him up against one of the most stubbornly held beliefs in AI research, which is that our knowledge of the world comes from models, mental images we build from observation and experience. Human beings work with models to interact with the world at large. And yet, Brooks noted, we can perform tasks, such as catching a Frisbee, which we don't have time to model or even think about. Yes, complex analytical tasks such as weather forecasting or stock trading require models, but the vast majority of day-to-day activities, such as walking or riding a bicycle, don't require us to think about the task at hand. We just *do* it.

In 1984, when Brooks arrived at MIT's AI lab, projects like Shakey were the norm. This faith in the supremacy of the model extended into the way AI researchers went about their work. Why build a physical robot when you could make a model of it inside the computer? The physical world, laden with mechanical problems, such as bad wiring and poorly machined components, made the difficult job of designing intelligent robots even more difficult, adding another layer of complexity onto an already overwhelming task. Most robots lived their lives as programs inside of computers, entirely virtual.

Brooks surmised that researchers were trying entirely too hard to make robots think like human beings. In fact, they were trying to make the robots outthink humans, saddling them with detailed minutiae that wouldn't even reach the conscious thoughts of a person. What if we build a robot that doesn't think? asked Brooks. While that seemed ridiculous, Brooks wanted to try an experiment, just to test his incredible proposition. Allen was the result; a simple robot proved that intelligent behavior didn't require a detailed model of the world.

To understand why Allen worked, why he could avoid objects in the real world, accurately and at human speeds, we'll need to look at how Brooks rewrote the rules. First, Brooks gave Allen a very thin layer of consciousness. Allen doesn't think about the room he's in, doesn't try to build a model of it or understand the shapes of its objects. Instead, Allen has been built with a very tight connection between his sensors, which detect the presence of nearby objects, and his affectors, which move him about on his bar-stool pedestal. There isn't a lot of fancy computer hardware between Allen's eyes and legs, as it were, to gum up Allen's thinking. It's as if Brooks programmed Allen into a Zen state of no-mindedness. That absence of reflection—sought after by spiritual adepts—gave Allen a clarity of purpose that allowed him to fulfill his mission: keeping away from things.

Second, and at least as important, Allen is a physical object that exists in the real world, not just in the memory of a computer. Brooks thought that embodiment was an essential detail in artificial intelligence—not an afterthought, but the main event. Without experience in the real world, Allen couldn't learn. Repeated interactions with the real world, such as hitting a wall or running into a person, would teach Allen how to behave. A narrow consciousness of the here and now, combined with a physical presence, would provide *everything* Allen needed to achieve his goals. And, as if Brooks needed further confirmation of his theories, Allen worked on the first try.

So Brooks built another. And another. And another. Pretty

soon all sorts of critters infested the hallways of the MIT AI Lab—gathering up empty soda cans, searching for power outlets to plug into and recharge, or just acting skittish and shy. Each of these simple robots outperformed competing projects that had taken years to design, while Brooks could whip up a new robot in a matter of weeks. As he got more comfortable with his designs, they grew smaller and sprouted legs.

Another of the major challenges confronting AI dealt with the problem of locomotion. Most creatures use four or six legs to move about. Humans use only two, and we know that a lot of our here-and-now body thinking focuses on maintaining our balance in midstride. Robots with legs had never been very successful, but unless scientists planned on restricting robots to paved environments, legs would be absolutely essential. With legs, you can climb over obstacles in the environment. Without them, you're stopped cold.

Genghis, like Allen, was both stupid and ugly. But he was tiny—under six inches tall—and had six legs. Each of these legs was independently controlled; each leg had no idea what its companions were doing. But surprisingly, Genghis learned to use his legs so well that pretty soon he could encounter obstacles, such as a wooden plank in his path, and spontaneously climb over them. This wasn't something that Brooks had programmed into Genghis; the behavior had *emerged* out of Genghis's interactions within his environment. Brooks got something that couldn't even be done in old AI out of Genghis, for free.

Although these emergent behaviors might seem almost magical, they were really the by-product of the robot *and* its environment. Earlier robots, programmed to understand everything, couldn't cope with changes in their environments. Brooks's simple robots would make mistakes, but each of these errors would temper their future behaviors. You couldn't program Brooks's robots; they had to learn for themselves.

Something was very right about this whole approach, but it

wasn't reasoned or rational, at least, not in the classical sense favored by AI pioneers. Brooks couldn't tell you exactly how Allen or Genghis worked—only that, if a creature was built a certain way, certain patterns of behavior would emerge. The proof of the approach was in its successes. Of course, by now, even venerable godfathers like Marvin Minsky had begun to rethink things a bit. In 1986, as classical AI faded away, Minsky published *Society of Mind*, which marked the beginning of the public revolution in artificial intelligence. In it, Minsky hypothesized that human behavior couldn't be described by a single influence, a "little man inside your head" who gave you meaning and motivation. Instead, he suggested that the thoughts you think and the things you do *emerged* from the competitive squabble between little minds within your greater mind. For example, in a given situation, such as a date, you might have romantic motivations, amorous feelings, and the delight of a specific activity (such as watching a movie). But that's only the short list of things you're conscious of. All of it is making demands on you, calling you to activity. Some of these voices inhibit others, some amplify others; when a voice gets loud enough, you'll act.

Minsky's ideas have been embodied in toys such as Tamagotchi and the Furby. The idea of competing needs generating realistic behavior has proven to be another of the major breakthroughs in artificial intelligence. When the AI community moved away from theories depending on the unity of consciousness and considered the self as an assemblage of fragments, it finally began to generate realistic simulations of behavior. This, along with Brooks's ideas about the thinness of consciousness and the necessity of embodiment, began to radiate out across the scientific community, and a "new" AI began to rise from the ashes of the old.

Fifty years ago, Norbert Weiner predicted that intelligent machines could be designed, provided we could build into them enough capacity to adapt. Rodney Brooks created the mechanism of adaptation, an embodied but uncomplicated system of sensors and affectors creating emergent behavior patterns. But would this work for creatures of great intelligence? Could an artificial consciousness similar to our own be grown from Brooks's designs? All of Brooks's robots, he freely admitted, had less intelligence than a cockroach. Much less. Could you span the three hundred million years of the evolution of the nervous system in a single step?

For a researcher, there's only one way to find out.

In the summer of 1993, Brooks embarked on his most ambitious project to date, one that will, if successful, bring him renown well outside the tight artificial intelligence community. Intent on realizing the dream of AI—a human-quality synthetic intelligence—Brooks is scaling his robots up. Way up.

CHILD'S PLAY

In the first half of this century, an enterprising Swiss scientist defined the field of childhood development—how children learn about the world, and how their thoughts form. Jean Piaget opened the world of the growing child to scientific study, and yet he never considered himself a developmental psychologist, even though, in a practical sense, he founded the field. Piaget's lifelong obsession centered around a study that became known as genetic epistemology—that is, how our genes and their expression in human biology create our understanding of the world.

Piaget's children were his own test subjects, and as they grew from infancy and into adulthood, Piaget watched them closely, as both father and scientist, carefully recording their words and actions as they played. He listened to them as their minds grappled with the reasons why things are the way they are. It's interesting to note that no one had done this, *ever.* Previously, it had been assumed that children began life as irrational beings, learning the logic of the world as they grew to adults. What Piaget uncovered—a discovery Albert Einstein called "so simple that only a genius could have thought of it"—was that children had their own way of knowing, a coherent and logical framework of thoughts which, while not correct from an adult's point of view, is not incorrect either.

In contradiction to the image of the child as illogical, Piaget proposed that children are constantly exploring the world with their senses, and that their interactions with the world help them to formulate theories to explain the activity of the world. At every step along the way, the child applies its growing base of logical thoughts, testing them for truth, amending them as new experiences force a reconsideration of their understanding. Their play in the world is actually an advanced experiment in the behavior of that world, and every interaction leaves the child with a broader and more complete sense of the real. Piaget believed that, as the child grows up, each piece of experiential knowledge becomes the foundation for further investigations, that the real world acknowledged by adults *emerges* from these basic experiences.

All of this should sound quite a bit like Rodney Brooks's approach to making intelligent robots: it's only through the robots' continuous play in the real world that they can develop the techniques for exploration of that world. This isn't an accident: Brooks and his team at MIT studied Piaget's work, and found within it a model that could potentially lead to a robot with human qualities of intelligence. In this sense, Brooks's robots are designed to be infantile so that they can grow into intelligence. Just as we do.

The centerpiece of Brooks's laboratory in Cambridge is a large mechanical man, all burnished steel, with exposed wire ribbons running from his torso into a collection of computer equipment. His head looks like a metal hoop onto which two pairs of cameras have been mounted. (Yes, this robot truly has four eyes.) His arms are fully jointed and terminate in metal fingers. If you slapped skin on him, he might not be a bad-looking fellow. Perhaps a bit strange, but not entirely out of the ordinary.

COG, as he is affectionately known, represents the summing-up of all of Rodney Brooks's research in artificial intelligence, the testing ground for a theory both radical and simple: can he create an infantile human consciousness from an artificial intelligence? Before humans grow up and possess all the skills that we believe make us human, we are fully equipped with senses and abilities that allow us to learn about the world, to learn *from* the world. So Brooks and his research associates have done their best to fit COG with sensors that duplicate—in part—the set we're blessed with from birth. COG can see and hear, can feel his body, can move his head, neck, arms, and torso in just the same ways that a human infant can. While far less than a perfect replica of a human infant, COG has enough capabilities to demonstrate some extraordinary qualities.

One of the very first things that an infant learns how to do is to use its eyes to track objects in its field of view. Biologists have shown that some of this ability is innate—that is, we're born with it—but that we need to learn how to use our eyes and neck muscles together to follow objects as they move outside our immediate focus. Infants have the need to do this (just as Allen has a need to move away from people), so they practice movements of both sets of muscles, progressively correcting their errors, until they have mastered that ability. Similarly, COG has been programmed with a thin consciousness coupling his eyes and his body into a simple system of reflexes that have allowed him to learn how to track the objects that he can see. Brooks didn't teach COG how to do this, but he built COG in a way that allows COG to learn from his own mistakes. COG will try a number of experimental movements until he finds the correct mixture of eye and neck motion allowing him to follow any object around him.

As we learned earlier, the human infant also has an innate

capacity to recognize faces, and a similar ability has been built into COG. This means that COG can respond to faces—those of people or of other objects with faces, such as toy dolls. Brooks believes that any significant breakthroughs with COG will come from social interactions—that is, interactions between COG and his "caregivers," as the researchers have fondly termed themselves. Recognition is the bedrock of social interaction; if a child can recognize a face, that child will begin to watch the face closely, interested and absorbed.

It is believed that mimesis—imitation—plays an important role in learning, and again, this appears to be another innate drive of the human infant. COG too has learned how to mimic the motions of other faces: that is, if a researcher nods, COG can duplicate the maneuver. This is more complicated than it sounds; COG has to observe his caregiver, using his eyes and his computer programs to extract the movements of his caregiver's face. COG then repeats the maneuver, observing himself as he tries to imitate the caregiver. COG might not get it right on the first try, but after a few attempts, he'll have the movement down cold. This is a task other robots have found very difficult to perform, even though humans can do it without a thought.

So far COG has displayed an interesting set of emergent behaviors. Demonstrating a skill far beyond other robots, COG can play with a Slinky, something he can do because he can rely on his internal sense of his body. Each of his joints is equipped with sensors to detect position and pressure, just as ours are. Without them, COG wouldn't be able to gauge his own movements as he played with the Slinky; with them, he can listen to his own body, using the feedback from his sensors to learn how to keep the Slinky in a perfectly smooth motion.

These are some neat tricks, but (and here's the $64,000 question) is COG conscious? In the sense of a human consciousness of language and objects, certainly not. However, he has some of the conscious capabilities of an infant—that is, he displays the same sorts of behaviors, and he acquires those behaviors just as an infant does—so, if there were a Turing test for week-old humans, COG might score in the neighborhood, which is quite a feat in itself.

Whether Brooks can truly re-create the consciousness of a human infant in COG's body is a question only time can answer. It might be a blind alley, and in twenty years others might be discussing the failure of the new AI. So far, the research is very promising, and we may yet be able to grow intelligence with human characteristics. But in order to do that, we'll need to look at another kind of behavior humans come equipped with—emotions.

We know that human beings have emotions from birth, though an infant clearly has a limited repertoire of emotional states—most obviously anger, fear, hunger, discomfort, and contentment. From this narrow base, the broad and subtle range of feeling emerges as the child grows older. We hold it as a hallmark of emotional maturity when an adolescent falls in love for the first time. But before this, children learn all of the other various emotional states, both positive and negative, from their experiences in the world. They experience emotions, such as disappointment, and learn how to assign a name to the feeling, so they can communicate their feelings to others. And, it is now believed, emotions actually interact with our thoughts; that is to say, we can't think or come to decisions *without* an emotional bias. (Mr. Spock aside, emotions are more than an irrational component of human being; they're integral to our ability to behave *as* human beings.)

So to be anything like a real human infant, COG must have emotions, along with emotional needs that he finds himself driven to satisfy. That drive will force COG to learn even more about how to be a real infant in the real world. It's from this observation that the worlds of the Furby and of COG, at opposite ends of the spectrum of robotic intelligence, begin to converge. The Furby *is* an emotional creature, with basic needs and desires. If the Furby can ape emotions, so can COG. But COG is a thinking creature whose behaviors evolve over time, like a person. Can a machine have emotions? And can a machine with emotions truly be called a machine?

3 THE CHARMED WORLD

GOD AND GOLEM

Two hundred years before The Turk astonished eighteenth-century Europe, legend has it that Rabbi Judah Löw ben Bezulel, the *maharal* of Prague, conducted the first successful experiment in artificial intelligence. Using the word of God as his guide, the rabbi took the clay of the Earth—just as God had done with Adam (whose name literally means "clay")—and formed it into the shape of a human being. A mystic whose teachings influenced Judaism significantly, Rabbi Löw inscribed certain Hebrew incantations—spells, if you will—onto the inert form, literally drawing them into the clay. Following the Creator's approach, he gave it *ruach* (spirit or breath), and it came to life!

This form endowed with life—a golem—became the servant of Rabbi Löw and the protector of the Jews of Prague, defending them against a gentile population who had regarded them with fear and suspicion. Once these bigots confronted the golem, they backed down and allowed the Jewish community to live in relative peace, leading to a brief golden age of Eastern European Judaism.

Different stories are told about the eventual fate of the golem. One has it that Rabbi Löw, who grew concerned at the golem's prodigious abilities, erased a single Hebrew character from the golem's skin, changing "breath" into "death," and so the golem reverted back to inanimate clay. Another holds that the golem remains alive to help the Jews in times of trouble. In any case, the story has become one of the most-repeated tales of medieval Jewish life, with an influence far beyond the ghettos of Eastern Europe.

I knew of the tale of the golem—it was even once featured in a striking episode of *The X Files*—but until I met Anne Foerst, I never suspected how much of modern artificial intelligence research has grown up around that myth. Marvin Minsky first heard the tale, traditionally passed down from grandfather to grandson, as a child. Minsky learned the magic of the incantation, a secret supposedly held only by the descendants of Rabbi Löw. He was not alone in this; many of the Jewish AI researchers at MIT had learned the tale from their grandfathers. This was a strange coincidence: could all of them really have been descendants of Rabbi Löw? Had this influenced Minsky's choice of career? Had a childhood encounter with the magic of artificial intelligence propelled him toward its scientific realization?

At a conference entitled "Science and the Spiritual Quest," Anne Foerst laid out a case to support this curious proposition. Rather than developing from strictly scientific principles, the drive of these AI researchers came partly from the faith that such a creation was possible—and these presumed descendants of Löw felt themselves uniquely qualified to resurrect the golem.

How could Foerst make such a bold statement, one likely to raise the ire of the scientifically minded AI community? Foerst had been working with AI researchers at MIT for several years and had gotten to know them well, coming to an understanding of how they thought and what forces drove them to their work. As part of her research at Harvard's Divinity School (Foerst is a cleric, a Lutheran minister), she did her field study with these scientists, in residence at the ultra-rational AI Lab, like some anthropologist living among a newly discovered Micronesian society.

Rather than being seen as extraneous to the labor of producing artificial intelligence, by the time she had finished the work on her thesis, Foerst was considered essential to it, and she accepted an appointment as the resident robot theologian at the lab. Her studies had pointed out several assumptions—articles of faith, so to speak—that had handicapped AI researchers, unspoken and sometime unconscious biases about the nature of mind and intelligence, biases that had subtly pervaded their work. Many of these assumptions had their roots in religious beliefs—fine for the church or synagogue or mosque, but not necessarily helpful in the laboratory. Foerst's examination of the scientific myths surrounding artificial intelligence had improved the quality of their research.

In truth, robots will not need theological counseling for quite some time, if ever. But those who work with these growing intelligences need to be acutely aware of their own behaviors. If COG, as a social machine, learns from his interactions with his caregivers, then they have a responsibility to conduct themselves as if COG were a living being, with the same delicate psychological structure as a human infant. This is something that hadn't been thought necessary before Brooks began the COG project, since a human-type consciousness had never been attained in artificial intelligence circles. Once that goal had been stated, *all* of human ethics came into play, and Foerst set out to make the lab a safe and healthy place for an emerging intelligence, much as any parent would.

Foerst counsels researchers working with COG, but her current focus is with another project, an offshoot of COG known as Kismet. As Brooks's team became aware of the role of emotions as foundations of social interaction, they examined ways to explore emotional expression and emotional interactivity with a caregiver. Most of an infant's social interactions, based in facial expressions, couldn't be duplicated by COG. COG has cameras for eyes and microphones for ears, just sensors without any of the affectors, such as a mouth or eyebrows, needed to engage a caregiver in an extended emotional conversation.

So Cynthia Breazeal, a graduate student working for Brooks, worked out the design of a robot specifically equipped for emotional facial interactions. Kismet has no body, just a head with Ping-Pong-shaped eyes, long, pointy ears, and a mouth that can adopt a number of meaningful positions. The whole thing—although steel and wire, and lacking fur—looks surprisingly like a Furby and is only just a little bigger than the playful pet. This actually makes a great deal of sense: both Furby and Kismet, designed for emotional interactions with caregivers, need to be able to pull the same heartstrings. And some of those heartstrings are hard-wired, based in facial expressions.

Moving its eyes, ears, and mouth in synchrony, Kismet is capable of a fairly broad array of facial expressions corresponding to emotional states: calm, anger, surprise, sadness, upset, confusion, disapproval, happiness, excitement, exhaustion, boredom, interest, shock, and depression. That's a nearly full repertoire for a human being; for a robot it's absolutely remarkable. But more remarkable still is how Kismet adopts these expressions.

Intensely aware of the surrounding world, Kismet has the ability to see and recognize faces, an ability inherited from big brother COG. Like COG, Kismet has been programmed with drives it must satisfy, but these drives aren't just the fulfillment of a task, such as the clever manipulation of a Slinky. Instead, Kismet has emotional drives—derived from the examination of infant humans—and these drives argue within the thin space of Kismet's artificial consciousness to produce emotional behavior and emotional reactions. As with Brooks's other robots, there's very little thinking going on inside of Kismet—the connection between its emotional drives and its emotional state is close to immediate—just as with human beings.

When Kismet is turned on, it enters a state of calm. But Kismet has a drive to explore, to interact—in particular, to interact with human beings. If it doesn't receive stimulation, it will grow bored, or perhaps even angry and frustrated with the situation. These emotions, as they come and go, are reflected in Kismet's facial expressions. Conversely, if a caregiver such as Breazeal sits before Kismet, it perks up immediately, becomes attentive and watchful. Kismet can track a caregiver's face, keeping it in view, looking for some stimulation. The caregiver can play games with Kismet, such as hiding from view and returning (peekaboo), and Kismet's reactions will closely parallel those of an infant human. The caregiver can present Kismet with toys, and Kismet, under normal circumstances, will respond with delight—for a while, anyway. Just as with human infants, too much of the same thing will bore Kismet, and it will become disinterested in the interaction.

Kismet also has the desire to feel safe—another desire innate in humans—and will react if it feels threatened. For example, a Slinky in gentle movement will fascinate Kismet, but should the undulations become too broad or come too close to Kismet's head, the robot will react with a growing expression of fear. Too much activity in a caregiver's interaction will overstimulate Kismet, and if there's no letup in the bombardment, Kismet will simply disconnect from the situation, close its eyes and try to pretend it's not happening, just as babies do. (And more than a few adults.)

Once again, it's not Kismet's programming that drives its emotional states; rather these states emerge from the information pouring in through its sensors. Just like a Furby, Kismet has needs that must be fulfilled; Kismet wants to be entertained, to play, to discover new things. When these needs are met, Kismet is happy. If these needs go unmet, or when it becomes overstimulated, Kismet becomes angry or sad. Like a human being, Kismet has a comfort zone in which it delights to operate, a delicate balance of interaction, novelty, and repetition that keeps it engaged and happy. Move out of that comfort zone and Kismet will immediately react, attempting to communicate its own emotional state to its caregiver, using nonverbal language to articulate its needs, just as an infant does.

Recent research at Ohio State University shows just how well infants can communicate with their mothers. A group of preverbal infants and their mothers were taught a simplified sign language, a pidgin not too different from that which Furby speaks. Although these infants couldn't voice their needs, they could use the sign language to make requests such as "give bottle" and "stop," indicating an innate sophistication in human communication that had been suspected but never demonstrated. We're born knowing the importance of communication and fully able to communicate—though it takes some time for our bodies to catch up. (Some researchers believe this means that sign language predates verbal language in human evolution.)

Kismet has an innate facial sign language that allows it to learn to express the way it feels. This is a first step into something much more profound—an artificial intelligence capable of a broader, evolving communication with a caregiver.

Which brings us back to Furby.

IT'S ALIVE!

If anyone understands the complex relationships between children and their interactive toys, it is Sherry Turkle. For twenty years, from her chair as professor of the sociology of science at MIT's Program in Science, Technology, and Society, she has researched and written about how children relate to technology, how they work with it, and how technology transforms the way they think. In 1984, Turkle published *The Second Self: Computers and the Human Spirit*, which took a hard look at the relationship between the human soul and the cold logic of the machine, finding a surprisingly human space co-evolving between them. Computers, she argued, acted as a mirror; much as we might peer into a looking glass to examine our own face, she found a generation of children and young adults using technology to explore and understand the dimensions of the self.

This might seem a bit hard to accept at first. After all, aren't computers first and foremost *tools*, designed to aid us in our daily work? Turkle probably wouldn't argue that, but she would point to a growing wealth of studies that show how the computer can engage us as intellectual partner, playmate, or artistic palette, types of interaction that demand self-expression by the very nature of their activity. We invest ourselves in our relationships with computers, and from this we learn more about ourselves. (If you think about how people can become very protective of their computers—anthropomorphizing them—it's not too hard to see her point.)

Like Piaget before her, Turkle watched children as they played with computers, and she'd ask them questions that helped her to build models of the child's own understanding of the relationship between human and machine. As the 1990s dawned and the Internet and the Web assumed a more important role in the computing experience, Turkle began to see another type of self-examination and self-identification emerge—a life lived online.

In 1994, Turkle published *Life on the Screen: Identity in the Age of the Internet*, her findings based principally upon interviews with users of MOOs. (MOO stands for multi-user dimensions, object-oriented, a clumsy name that refers to the programming environment used to create MOOs.) MOOs are very similar to today's more familiar online chat environments, but unlike the featureless chat rooms offered by AOL, MOOs have programmable qualities, so they can be completely designed by their users to be a reflection of a user's identity. Several years ago I created my own room in a MOO; to do this, I provided a text description of the room, just a few sentences that would give visitors to my room an understanding of where they were and what it looked like. All objects in a MOO have descriptions, and you can create nearly any description you care to dream up. I then populated my room with objects, such as my chair and my desk. Visitors to my room could sit in the chair or read papers that might be on my desk, just as they could in the real world. But it was all done literally, with text standing in for real-world objects.

Although they are text-only environments, a well-constructed MOO very nearly reads like a good novel, with plenty of narrative to accompany the visitor in a journey through the space. Case in point: A well-known MOO run out of MIT's Media Lab (MediaMOO) describes a virtual duplicate of the Media Lab itself, and you could find within it many of the physical offices, meeting rooms, and staff members that existed in the real world. Several of the students at the lab kept a virtual presence within MediaMOO, participating in both the real-world activities of the lab and the online lectures, presentations, and informal discussions featured in MediaMOO. (To do this, they'd need to be in front of their computers nearly continuously, not an unusual state of affairs at MIT.)

But not all MOOs were so professional or so close to a real-world counterpart. MOOs allowed people to dream up their own universes, to become characters in a play of their own design. This, for Turkle, was another important vehicle for self-expression and self-definition in the world of the computer. She found a community of tens of thousands of individuals who had re-created themselves, building a universe to match, and wanted to understand how they could juggle all of these separate identities. She found that MOO users would often exaggerate characteristics they lacked in the real world; for example, shy individuals might become effusively extroverted. Turkel found that experience in MOOs helped people to rehearse social behaviors they found difficult in real life, such as talking to members of the opposite sex. MOOs helped many people to explore sides of themselves that they might not otherwise have access to.

A famous cartoon in *The New Yorker* shows a canine, computer mouse in hand, chatting with another four-footed friend.

His line? "On the Internet, no one knows you're a dog." In the same way, gender bending became the unspoken pastime of many MOO users, who explored alternate sexual identities or sexual roles—behaviors strictly taboo in the real world, but accessible through the anonymity and freedom of the MOO. Famously, Turkle likened many of these behaviors to those seen in rare cases of disassociative identity disorder (DID, once known as multiple personality disorder), for as her study subjects "stepped into" their online personas, they would *become* that persona. Of course, these two are quite different. The soul afflicted with DID doesn't have control over the comings and goings of the fragmented selves that make up his or her personality, while MOO and chat room users can effortlessly slide between personas, even managing several of them simultaneously. This radical act of self-definition, Turkle argues, is the hallmark of the relationship between man and machine.

Lately, Turkle has been studying Furbys.

One of the questions of childhood development that Piaget chose to address concerned the idea of the animate object. Human beings have very little difficulty determining whether objects in the world are living or dead, but we're not born with this knowledge; it's acquired as we mature. Piaget learned that a child initially makes no distinction between animate and inanimate objects. Everything, in some sense, is equally alive. With that theory in hand, the child goes on to discover that all things aren't equally animate. A toy truck, for example, is animate only when pushed, only when the child moves it; from this the child begins to make a distinction: objects that are animate don't require any action to activate them. They're active all on their own.

This seems to be knowledge about the world that we acquire between the ages of four and six, important information that helps us to set up our expectations about how the world will behave. (If we're not activating it, the inanimate world won't change.) For all of human history, adults have neatly divided their world into two parts: the animate and inanimate, the living and the dead. This may now be changing because of the Furby.

A page on Turkle's website at MIT issues a call for children who own virtual pets. Turkle is collecting stories from them, gathering their experiences, and what she's learning is as startling as anything she's discovered before. These children, interacting and relating to these virtual pets, are creating a *third* answer to the question "Is it animate or inanimate?" It seems that the distinction between living and nonliving, which used to be so very clear, has become muddy. And as these children learn and play within the world, this confusion seems to persist. They don't want to class the Furby as an inanimate object, *even though it is a machine,* because it shares some of the qualities of fully animate beings. So rather than throw the Furby into one category or another, these child-philosophers have opted for a novel approach: they're creating a new definition for partially animate objects. Certainly not human or even animalistic, but animate just the same.

This is where Furby differs from all of the toys that have preceded it. None of them forced a redefinition of the categories we use to make sense of and give order to the world. The Furby is not an exception, but rather, a new class of toy, which, by virtue of its capacity to mimic human relationships, has created a new standard by which we distinguish the living from the dead, human from machine. And the Furby is only the first of its class.

PRIVATE LANGUAGES, PART I

The future can be a very funny place, particularly when it arrives ahead of schedule. Consider a scene from Woody Allen's fast-forward science fiction farce *Sleeper*, when he encounters the twenty-second-century version of man's best friend. The furry but obviously mechanical pet presents itself and says, "Woof!

Woof! Woof! Hello, I'm Rags. Woof! Woof!" Allen pauses and then returns: "Is he housebroken or will he leave batteries all over the floor?"

We may never know if the designers at Sony had seen *Sleeper* when they developed AiBO. Introduced to the media in May 1999, the world's first electronic dog caught the public's attention—and pocketbook. About six inches tall, a foot long, and looking very much like an immature aluminum dachshund, the creature is cute and playful. At $2,500, AiBO costs a bit more than a Furby (about a hundred times more), but AiBO is more than just a scaled-up Furby. It's the sign of things to come.

AiBO's name comes from "Artificial Intelligence roBOt"—but it also means "pal" in Japanese—and it's clearly designed to be a companion, like the Furby. Unlike the Furby, AiBO has a lot of expensive computing power (hence its price tag) and a broad set of capabilities. The Furby reveals its capabilities as it grows up—but these qualities aren't learned so much as released, in a step-by-step process. Borrowing from research done at the MIT AI Lab (which raised a few eyebrows in Brooks's lab, as they weren't consulted on AiBO's design), AiBO learns how to stand up, walk, and do tricks as it matures. Its life cycle is broken into four stages, like the Furby: toddler, child, young adult, and adult. Like the typical adolescent, AiBO is hardest to handle in the young adult phase, but will eventually begin acting like a responsible pet when it's fully grown.

Some of this must be sleight of hand (otherwise AiBO would be the greatest achievement in the history of robotics), but some of it seems real, particularly in how it learns to walk by trial and error. AiBO uses sensors to read its own body, just as COG does, learning through feedback how to move itself successfully through the world. Granted an insatiable curiosity in its youngest days, AiBO uses that drive to improve its interactions with the surrounding world—just as COG and Kismet do.

When Sony offered a limited supply of AiBOs for adoption, customers practically fell over each other trying to acquire one—something else AiBO has in common with the Furby. The three thousand intended for Japan sold out in twenty minutes, while the ever so slightly more reluctant Americans snapped up two thousand in just four days. Given that AiBO costs more than a good computer system, it seems likely that the "kids" buying the toy were actually full-grown (and wealthy) adults.

And AiBO is only the tip of the iceberg. The success of the Furby has left the toy industry scrambling to improve on something that has already conquered the world. So far, Hasbro has partnered with Rodney Brooks's start-up venture, iRobot, to create My Real Baby, which promises that it "cries and coos, instantly responds and reacts, smiles and squeals with delight." Brooks is taking the intelligence of COG and Kismet and putting them into a toy shaped like a human baby. My Real Baby and its successors will make the Furby look like a Stone Age axe next to a spacecraft. But Caleb Chung, creator of the Furby, isn't resting on his laurels. He's partnered with Mattel to create Miracle Moves Baby, using voice recognition and touch sensors to create a toy that can respond to verbal commands. Both toys have internal clocks so that the dolls behave like real infants, demanding to be fed, played with, and put to bed on a regular schedule.

When I began working on this book, I assumed that Kismet would become the foundation for a next-generation Furby. But when I presented that idea to Anne Foerst, she reacted violently. "Kismet *needs* a caregiver. The problem is that a child's response to Kismet might be wrong." Given that children are still learning about the world themselves, this seems more than likely. Foerst asked me to consider a house pet. Many children get puppies or kittens, but most children don't raise them by themselves until they're seen as mature enough to handle the responsibility. And we've all seen a dog or cat raised by a careless, unthinking owner—an animal psychologically damaged by its upbringing.

Could something similar happen to Kismet if it found itself in the body of a Furby and in the hands of an unthinking child? "Without question," Foerst replies. "It's up to the caregiver to give Kismet social responsibility," which means that Kismet needs an example of good behavior if it is to express such behavior as it matures. Just as people do. In some sense, Kismet is too *alive* to be considered a toy, too sensitive to the world around to be simply plopped into the arms of a growing child. Yet there likely is a middle path, and this path shows us how profoundly the relationships between humans and machines will change over the coming years.

Consider again our child, born at the turn of the millennium, as she wakes on her sixth Christmas to find "the hottest toy in the world" under the tree. It's a very special present, a "virtual friend" of the latest variety. She unwraps it and—with the help of her parents—puts the batteries in and turns it on. But she's told that she can spend only an hour a day with her new friend, and that Mommy or Daddy has to be with her during that time. The toy is practically a newborn, and this confuses the child—she thinks it might be broken—but Mommy tells her that she was once like this herself, and with a little love and care, her new friend will grow up to be a treasured companion.

So the three spend an hour together—daughter, mother, and toy—and each learns how to respond to the other. The child learns that she mustn't get cross with the toy, that the toy will get upset if it is yelled at, or sad and depressed if it is left alone too long. But already the child is learning how to read the toy's many expressions; already there's a bond of communication between child and toy that will endure and deepen as days turn into weeks, then months. By her seventh Christmas, the child thinks of that toy as another companion, a real being just like her parents or her kitty, and the toy has evolved a wide range of behaviors. It can follow the child around the house, explore, listen to stories, and, just now, begin to recognize objects.

From the beginning, the child has talked to the toy, endlessly repeating the trivia of childhood, but sometimes pointing out specific objects with specific words: chair, table, crayon. After a long period of interaction, the toy begins to recognize these on its own and begins to respond with its own bits of language, just nouns free of any particular meaning except the objects themselves. It's not much more than the announcement of recognition, but it's still important, and the child praises the toy for its accomplishment. The toy responds with gratitude.

As time passes, Mommy gives her daughter more independence as she plays with the toy; the ground rules have been laid out, and the child has begun to understand that the toy has feelings—just as she does—that can be hurt. In a way, the toy has become the child's tool for understanding her own feelings. They're both sensitive in precisely the same ways, each a reflection of the other's behaviors.

As the child grows, the toy grows, too. At one point, Daddy takes it away for a new body. His daughter cries—she's never been apart from it for long—but he reminds her that just as she is growing taller, so her toy must grow. He comes back with the toy in a shiny new body, a body that must learn how to walk again. So his daughter spends precious hours helping her friend learn her old skills in a new form, praising her for her successes and helping her through her mistakes.

A toy like this, if it ever could exist, would be more than an electronic playmate; it could help us raise better, more caring people from the infinite possibilities of childhood. When children can see their reflection, when they can come to explore and model their own behaviors on the outside world, they'll develop the theories they need to become better adults—more sympathy, more understanding. Certainly, this should happen in households with multiple children, but sibling rivalry presents a constant barrier to such deep interaction; children are in competition for the affections of their parents (at least, that's how they see it), and altruism is in short supply. But with such a toy, with such a mirror, a child could see herself objectively, could develop the qualities needed to cultivate another being's consciousness.

By the age of ten these two have spent four years together, growing up and nurturing a friendship tighter than most. In fact, from this toy the child learns what is most important in friendship: virtues such as loyalty, sensitivity, and trust. She'll know to seek these qualities in her other relationships. But more than this, the two have developed a complex secret language of gestures and words, so that each knows what the other is thinking without the benefit of many words. An artificial intelligence has been nurtured in the image of a human child, and a human being has been nurtured into full humanity by an artificial intelligence.

II

ACTIVITY

4 TINKER'S TOYS

HIP DEEP

"Click the play button *now*."

Click.

"To give instructions to your robots you will use a programming language called RCX Code. After creating your program, you will download it to your RCX. The RCX is the brain of your robot.

"To begin, open to pages 13 and 14 and build the Pathfinder One robot . . ."

My computer is talking to me. Not altogether rare in an age of interactive media. And, because it has encouraged me deep into a complex task, it gives me detailed instructions, encouraging words, and helpful images, doing its best to aid me.

The hardest part of the whole operation seems to be finding the correct pieces to complete the job. There are hundreds of little plastic parts (nearly a thousand!) in shiny grays, blacks, yellows, whites, and greens. It's all a little daunting, but eventually I realize I simply have to dunk for parts, diving hand-first into a dozen trays, bringing up a clutch to be eyed and sorted for the proper components. It takes a few minutes—in fact, longer than I really wanted to spend on the task—but as I progress, I begin to remember how much fun I used to have doing this.

As you can probably tell by now, I was one of those kids who loved Legos. Together with Hot Wheels and Tonka trucks, Legos made up a lot of the imaginary play space of my childhood. I spent hours deep in the construction of fantastic shapes: spaceships and houses and a whole host of other things I couldn't name. But I could take my imaginings and make them concrete, one brick at a time.

I always preferred the generalized Lego sets to the custom kits with the pieces for an automobile or a firehouse. Perhaps I wanted to discover the designs for myself. Creativity seemed to be the whole point, more involving than the assembly of some predesigned product. But now I find myself seated in front of a very specialized kit of Legos. This set will allow me to build *my own robots*.

All of this talk about robots and intelligence has gotten me itching to build a robot of my own. During the research phase of this book I came to understand that artificial intelligence is a sport that nearly anyone can dabble in—if you have the right equipment, that is. Now I've realized that I do indeed have the right equipment here in these boxes, potential resting in a multitude of plastic pieces. Cool.

Eventually, I do find all of the correct parts and snap them together according to the detailed photographs given in the *Constructopedia*. First the frame, then two motors, right and left, then the wheels—yellow hubs popping into rubber tires—gently pushed onto the shafts protruding from the motors. Now I add the power cables, which look like a pair of thin 2×2 Lego squares with a wire running between them, attaching them to the power plugs atop the motors. For the final stage, I take the RCX, the brains of my robot, and place it on top of the body I've constructed. It's about four inches on a side, and weighs half a pound, so it takes a moment of wiggling before I hear it snap into place. Now I take the unattached end of the motor power cables and attach them to the RCX, onto squares labeled "A" and "C." Okay, I'm done. But what does it do? What have I built?

"In training mission one, you will create a program to make your robot move away from you for one second, and stop.

"You will create programs by stacking code blocks that fit together like puzzle pieces. On the left side of the screen you will see four different-colored blocks. Each block is a menu. The Commands menu contains blocks to tell your RCX to turn motors on and off. Sensor Watchers contains blocks to tell your RCX how to respond to touch and light . . ."

I'm beginning to get the picture now. It seems that the Lego metaphor of bricks snapping onto bricks has been extended into the bits of software that will make my robot strut its stuff. This idea of component software has intrigued computer programmers for many years, but has only rarely resulted in anything usable. Most programs are too different, one from another, to be composed out of a heap of basic elements. A well-written program is like a poem, each element in a precise position that creates, in the eyes of other programmers, its essential elegance.

Such elegance is rarely learned in a week or a month or even a year. Most practitioners of the software arts take five to ten years before they've really mastered their craft—very much like the medieval concept of a trade guild and its stages of apprentice, journeyman, and master craftsman. Much of that time is spent learning the pitfalls of programming, by falling into numerous logical holes and endless loops. It's not that software is intrinsically hard, but it can get complex quickly. You really need to think about the problem before you begin. And this is what separates the seasoned programmer from the neophyte: experts think long and program little, while beginners do just the opposite, hacking their way to a solution, like an explorer clearing a jungle path with a machete.

You have to be pretty dedicated (or pretty crazy) to master computer programming, so most people shy away from it, a bit of "rocket science" many doubt they could master. But anyone can play with Legos. They're obvious, even to a child. And that's the secret of this software. Using the metaphor of the Lego brick, people should feel right at home snapping a program together.

A few moments later, I have snapped together three green virtual blocks on my screen, which I grabbed from the Controls menu, like fishing Legos from a bin. The first is labeled "On," the second, "Wait," the third, "Off," all stacked from top to bottom, just like a tiny tower. Of course, this is only a representation of the program. I have no idea what commands my Legos actually understand, nor do I have any need to know. But I do need to get these commands into my robot, so I press the button labeled "Download" with my mouse and move my little robot near the four-inch tall black-and-gray tower which I have connected to a port on the back of my computer. A green light glows from the tower, which tells me it's using infrared signals to talk to the robot, much as the Furby uses them to talk to its peers. Then my robot emits a melodic beep, and I know we're ready to roll. Literally.

I place the toy on the floor beneath me and press its green Run button. Voilà! The robot backs away from me for a second, then comes to a stop. In the parlance of these Legos, I turned its motors on, waited a second, then turned them off. That's the meaning of those three green blocks on my screen, snapped together into a program.

That's pretty cool, but hey, I've been programming for twenty years, so I immediately get more ambitious. I realize that I can control each of the motors independently: I can turn the left motor on while leaving the right motor off, so one wheel will spin while the other remains still. If I do that, my robot will turn. I could even alternate that motion so that my robot would slither along, turning first to one side, then to the other, like a serpent. And if I do it quickly enough, it will look like it's doing the jitterbug.

I look through all of the Command blocks and figure out which ones I need. Soon I have a neat little stack of blocks nestled within a larger pair of red Stack Controller blocks, which tell my green blocks to repeat their actions forever. I've created a single jitterbug movement, but those red blocks will cause my robot to reproduce that motion ceaselessly.

Another download, and I've got a slithering, jitterbugging robot, weaving its way across my carpeting. This *is* cool, and easier than I thought it might be. I start to scavenge through the kit, looking for other things to attach to my robot. I come up with a touch sensor, looking like a 2×3 Lego, but with a small yellow nib on its front, a switch that I can depress with my finger.

I gently pry apart my robot and add the touch sensor, like some ultramodern hood ornament, using another of the connectors to wire it into the RCX, to the position labeled "1." There's a particular Stack Controller block called "repeat while," which looks at the touch sensor periodically, and if the sensor hasn't been depressed, will continue to run the robot through its dance. On the other hand, if the touch sensor has been depressed—meaning my robot has hit the wall or some other kind of obstruction—the robot will simply stop in its tracks. Now that I've added a sensor to the robot, I can use that information, feed it back into its program, and get it to change the behavior of its affectors.

My knowledge of Lego programming is far from perfect, so it takes a few tries, but soon enough, my critter careens back and forth across the floor, comes to a wall—and stops. Another successful experiment. Well, why stop there? I suppose I could use that little bit of sensor information to reverse the direction of the motors, or I could even run one motor for a bit so that the robot turns around . . .

That's when I realize that I'm in hip-deep and loving it. Something about this toy is bringing out the kid in me, bringing me back to the creative child who loved to play with Legos. But I know a lot more than I did as a child and am fresh from a few weeks studying Rodney Brooks's robots.

I begin to wonder how hard it might be to build a Lego version of Allen, Brooks's first robot. All it has to do is run away from people. And that's not very hard. Hmm. A home-brew robot built out of Legos displaying the qualities of emergent intelligence? Now *that* would be cool. And I've got everything I need to make it happen.

THE (OTHER) HOTTEST TOY IN THE WORLD

Although it received just a tiny portion of the attention given the Furby, this toy I've been playing with, Lego's Mindstorms, was in very short supply during the Christmas season of 1998. At $225 a set, it shouldn't have been flying off the shelves, but FAO Schwarz, the Tiffany's of toy retailers, couldn't keep it in stock. Lego, no stranger to runs on Christmas toys, had been caught completely unprepared, and doubled their production volume to meet demand. The retailers cried more, more.

In 1958 a tiny company which took its name from the Danish for "play well" introduced a revolutionary product, an interlocking system of building bricks that allowed for an infinity of possible combinations. Lego (*leg godt*) had been born. By 1966, about the time I first began to play with these plastic parts, nearly a billion bricks a year were being manufactured, packaged, and distributed all over the world.

Legos joined a long tradition of similar toys: Tinkertoys, Lincoln Logs (invented by a son of architect Frank Lloyd Wright), and for the electromechanically inclined, Erector sets, with motors, gears, and pulleys. Yet each of these lost some of their luster once the genius of the Lego system came into full prominence. Something about these interchangeable parts seemed so essential, so basic, and so adaptable that they quickly became the creative medium of choice for children around the world.

Although it had taken Lego's toy designers two decades to perfect their plastic bricks, the toys proved practically perfect, and very little changed in Lego-land for forty years. They added a few basic components, such as the wheel, and extended the forms to include Lego figures, Lego families, and Lego astronauts, but it all remained variations on a theme, superfluous notes in an already perfect fugue. Children didn't need anything but the bricks themselves—preferably, a lot of them. (Given how long Legos last, there are likely many more Legos in the world today than people!)

In the late 1960s, Lego introduced Duplo, big blocks for children with tiny fingers, and scored a hit with the under-five set. Lego went after older children and young adolescents a few years later, in its Technic series, complex models with working motors that retained their snap-together ease of construction, but offered activity in equal proportion to imagination. Each of these had a fair amount of success, but nothing to compare with the original; the basic bricks continued to sell in ever-increasing quantities.

By the time the high-tech revolution was in full swing, as computers began to show up at home and in schools, Lego, working with a group of researchers at MIT, introduced Technic Computer Control, a system that allowed Technic robots to be controlled via a computer. Distributed only to schools in Denmark and England, the product was more of a test than a toy, setting the stage for some pivotal mass-market product introductions in the 1990s, beginning with the Technic Control Center, a kit of programmable robots for the Technic series of toys. These early Lego efforts got some attention in the press and had a few buyers in the educational markets, but never really made it big. The programming, still a bit cumbersome, presented a barrier to all but the most inventive and intelligent children.

Lego, learning from its successes (and its failures), continued to refine its designs for robotic toys, with plenty of help from MIT. By 1997, the designers had gotten it just about perfect, and in the middle of 1998, Lego introduced the Mindstorms Robotics Invention System.

Lego never knew what hit them. Somehow both the product

design and the zeitgeist conspired to create a buzz around this newest addition to their product family, and they optimistically projected sales of 40,000 units over the Christmas season—at a hefty $225 retail price. But as September melted into October, Lego found itself swamped with orders. Toy retailers were restocking their supplies over and over. Every time the sets shipped, they immediately sold out. So Lego doubled their projections again—and still fell short. In all, over 100,000 Mindstorms sets sold before Christmas 1998. They had another hit on their hands.

As buyers mailed in their warranty cards and Lego's marketing department carefully studied the age and gender distribution of their customer base, they found something utterly unexpected. Half of the Mindstorms sets had been purchased by adults—for their own use. This was unprecedented in Lego-land: after all, Lego makes toys for children. Mindstorms had touched the child in a multitude of adults.

With its 727 pieces, including two motors and a few sensors, the Mindstorms kit differed from earlier Lego products in one significant way: it included a device known as the RCX, the Robot Control and eXecution unit, the brains of the beast. The top of the RCX has an LCD screen that displays the status of the unit, as well as pads where up to three motors and three sensors can be simultaneously attached. The motors can be turned on and off by the RCX—literally, it functions as a power switch—while the sensors can have their values read in and interpreted by a user-created program running in the RCX.

As with the Furby, another enterprising engineer has taken a look under the hood of the unit and documented his findings on the RCX Internals website. Like the Furby, the RCX contains a single printed circuit board with a variety of capacitors and resistors, the building blocks of most electric systems, embedded around a single large chip.

That chip, a microcomputer manufactured by Japanese electronics giant Hitachi, contains the programming used by the RCX to interpret commands sent from the computer: to turn on a motor or read a sensor or pause a moment. The RCX translates these commands into computer instructions required for the task. This means that the RCX can be given a series of very simple orders spelled out in a stack of virtual Lego bricks and then translate these into complex actions. It also means that other engineers far outside the influence of Lego can write software for the RCX using a variety of tools, not just those supplied with Mindstorms. And many have.

Almost immediately after the release of Mindstorms, a whole host of websites sprang up designed for the adult owners of the toy. Many of these sites took a sophisticated approach to the playful possibilities offered, demonstrating a plethora of different Lego robots that could be built out of the kit, along with their control programs. One owner designed a Lego sorting machine that could winnow through different-sized Lego bricks; another built a remote-control robotic arm, and yet another constructed a Lego race car. Some folks went even further.

David Baum, an engineer at Motorola, created a new programming language for the RCX loosely based upon the C programming language favored by the vast majority of software engineers. (Developed in 1974 at Bell Labs, today nearly all computer programs are written in C or C++, its younger brother.) Baum called his creation NQC, which stands for "Not Quite C," and it opened the world of Mindstorms to the professional programmer who felt constrained by the comparatively easy-to-use programming tools offered by Lego. Programmers, like race car drivers, often try to get as close to the metal as possible, equating power and performance with access to the internal guts of a computer.

This was no small job. As a corporate outsider, Baum had to learn how the RCX worked without the benefit of the technical information which would have made the job only moderately difficult. Instead, he had to play with the RCX for many hours, testing its capabilities, treating it like a black box and watching it closely as he drove it through its paces. From these observations he was able to duplicate the function of the software tools supplied by Lego, wrapped within a software development environment (NQC) that professional engineers would find more comfortable.

Why would anyone spend so much time studying a toy? Not for the money. Baum gives away NQC freely on his website, without even asking for so much as a thank-you. In the era of the Internet start-up and the IPO, with seventy thousand potential customers out there, this may seem a bit curious, but some things are done for love, and Baum is clearly in love with his Mindstorms. First and foremost, he's a robot enthusiast, and he wants to share his passion with the world. There isn't a whole lot you can do in NQC that you can't do with Lego's tools, but programmers tend to think of their favorite programming languages in much the same way you might regard a worn-in pair of shoes. The fit is comfortable, reliable, and friendly, and that means a lot when a programmer gets down to the mental gymnastics of writing software.

Baum is not alone. Other programmers have created RCX versions of FORTH, an old and venerable language with an intense community of devotees; PERL, a language often used to process web-based forms; and Java, the current hot computer language. If you know a computer language, chances are you can find a version to run on your RCX. And if you don't, Lego has created a programming environment even a child can use. In fact, it was meant for children, designed by a researcher who fully understands the importance of play.

A KINDERGARTEN FOR PH.D.S

Throughout the middle of the twentieth century, Jean Piaget, while developing the principles of constructivism—the idea that children learn about the world through discovery and experimentation—played host to a wide range of students and researchers. Through his mentoring, Piaget extended the influence of his ideas far beyond his Geneva home, and through one particular postulant became a fundamental influence in artificial intelligence research.

Seymour Papert began his studies under Piaget in the late 1950s, toward the end of the elder researcher's career. The interaction proved to be explosive for Papert, leading to a series of ideas that would fundamentally influence education. But rather than becoming an educator himself, Papert joined colleague and friend Marvin Minsky to found the MIT Artificial Intelligence Laboratory. Papert believed that Piaget's research would have broad applicability in the field, a theory AI researcher Rodney Brooks would put into practice in the 1980s when he built robots that grew into intelligent behavior.

In the mid-1960s, working with researchers at Bolt, Beranek, and Newman, a Cambridge think tank simultaneously working on the basic foundations of the Internet, Papert created a computer programming language that encapsulated the constructivist ideas he had learned in Geneva. The result, LOGO, was the first computer language designed to be taught to and used by children.

LOGO, originally envisioned as an offshoot of LISP, the programming language of choice for AI researchers, provided the child with a basic set of operations, such as adding two numbers and printing the result on the display, but LOGO also allowed the child to experiment with these operations, putting them together into novel combinations, which would then become parts of more sophisticated programs. Children, according to Piaget, build an understanding of the world from basic concepts that become the basis for more advanced ideas. LOGO worked along precisely the same lines: once a child perfected a set of operations through experimentation, these theories could be put into practice. With

LOGO, children could enlarge their understanding of the computer through their own idiosyncratic investigations, and, like a Lego tower, place newer elements upon older principles until something truly novel emerged.

Papert tested his creation with students in Brookline, an affluent community just a few miles up the Charles River from MIT, and realized that children learn more quickly when they have real-world objects to play with as they are learning. To this end, Papert developed the Turtle, a small, squat robotic creature that rolled about the floor and could be controlled with LOGO programs. This physical interface device took the abstract language of the computer and translated it into the robot's physical movements, a powerful amplifier of LOGO's capabilities. When children saw the results of their experiments reflected in the Turtle, they learned faster, because the Turtle acted as the embodiment of their thinking, a reflection of their growing understanding.

The little Turtle was both expensive and delicate, and as the 1970s progressed, the Turtle became virtual, a critter on a computer monitor under LOGO's control. Although no longer embodied, the on-screen Turtle became a key element in the LOGO environment as the personal computer revolution exploded in the late 1970s, creating a base for LOGO to move from the classroom and into the home. Two home computers became the platform of choice for home versions of LOGO. One of them, the TI 99/4A, developed by Texas Instruments, creators of the Speak and Spell, carried a $200 price tag—well within the reach of most middle-class families. The LOGO cartridge for the TI 99/4A opened the world of computer programming to a generation. Many of America's best programmers got their start playing with LOGO on the TI 99/4A.

The more successful platform for LOGO was the incredibly influential Apple II. Designed by Steve Wozniak, the wizard of Silicon Valley and cofounder (with Steve Jobs) of Apple Computer, the Apple II found a place in millions of American homes, and because of Apple's aggressive marketing to educational institutions, in many schools. Millions of schoolchildren got some exposure to LOGO through the Apple II, using Turtle Graphics to program their own artistic and scientific creations, learning by playing.

Papert published the results of his revolutionary work in 1980, in a book entitled *Mindstorms: Children, Computers, and Powerful Ideas*. Yes, the Lego kit is named in honor of his pioneering vision: children can master logical thinking and complex concepts through play and exploration. Papert's book influenced both educators and computer scientists, who came to see the computer as a vehicle for learning, rather than just an instructional engine. Before Papert, computers were frequently seen as drill instructors, pushing students through rote problems; after *Mindstorms*, the computer became a sandbox of sorts, flexible enough to adapt to the child's evolving imagination of the world.

Just as *Mindstorms* reached bookstores, the Massachusetts Institute of Technology, Papert's home base, began an ambitious project to bring itself into a leadership role in the twenty-first century. In 1980, Jerome Wiesner, former science adviser to President Kennedy, was nearing the end of his term as president of MIT. Although MIT's reputation had been made in the hard sciences, Wiesner recognized that its future belonged to visionaries who could translate these technologies into useful applications. This idea put him at odds with an older generation of MIT professors who disdained anything but the most rational and scientific avenues of research. But Wiesner found a kindred spirit, someone to help him realize his vision, in a rising star at the Center for Advanced Visual Studies—the closest thing MIT had to a graduate program in the arts.

Long before he became the poster boy for the digerati, Nicholas Negroponte looked into the future and decided that it would be digital and interactive. Although the Center for Advanced Visual Studies had been ignored by most of MIT's faculty as too fluffy, Negroponte leveraged the substantial technical resources of the Institute to create a hugely influential work of interactive media. Known as the Project ArchMach (Architecture Machine), the device consisted of a high-backed office chair placed before a large projection screen. Mounted on the armrests of the chair, a number of input devices—such as a keyboard and mouse—allowed users to manipulate a virtual office represented on the display. Negroponte worked hard to create an impressively realistic representation of a real office, with file cabinets that could be opened, containing files that could be placed onto a desktop and manipulated. Designed to be an idealized electronic work environment, the ArchMach pushed computer graphics—just at their own beginnings—to the limits of the possible, providing one of the first examples of virtual reality nearly a decade before VR pioneer Jaron Lanier coined the term.

In Negroponte, Wiesner had found a scientist who understood both technology and its application. They began to plan MIT's entrance into the softer world of applied design, an effort that became known as the Media Lab. Now world-famous for its pioneering research, the Media Lab has fulfilled Wiesner's ambition: a meeting of science and art in an environment where each informs the other. Of course, Negroponte couldn't create a world-class institution by himself. He needed a lot of help, so he quickly invited Seymour Papert to set up shop under the Media Lab's roof.

The Epistemology and Learning Group at the Media Lab, founded by Papert in 1981, clearly gives a nod to Piaget's concept of genetic epistemology, the idea that experience in the world creates the foundation of learning. It also reflects Papert's extensive experience in using computers as tools for discovery and exploration. Given his considerable reputation, the group quickly became a magnet for graduate students wishing to study under the godfather of educational computing.

As Negroponte set up the Media Lab on the East Coast, a high-technology industry began to flourish on the West Coast. *Business Week* sent a young reporter named Mitchel Resnick to California's Silicon Valley, where companies like Apple, Intel, and Hewlett-Packard dominated the emerging high-technology industry. A thousand garage start-ups flourished across the cities of the Valley, and Resnick covered all of it, taking the techno-speak of the Valley's nerd-kings and translating it into language a lay readership could understand. He saw himself as a bit of an educator, out on the front lines, bringing the business practices of the Information Age to a wider public, acting as a bridge between the well-established world of atoms and the nascent world of bits.

As Resnick spent time in this technological playground, he grew more and more excited about the possibilities latent in the new toys being created by the ever-increasing waves of engineering talent and entrepreneurial spirit flooding the region. In particular, Resnick was fascinated by how computers could be used to help people understand complex systems. Could they become a platform for explanation and exploration? As a reporter charged with explaining the complex, Resnick found himself irresistibly drawn to this aspect of computing, and after five years on the beat, he left *Business Week* to enroll as a graduate student at MIT, entering its renowned computer science program.

Early on, Resnick became friends with another graduate student, Steve Ocko, who shared his interest in educational computing, and a few months later, the two took a course taught by Seymour Papert, who filled their imaginations with the possibilities explored in the recently published *Mindstorms*. In another seminar, led by researcher Sherry Turkle (soon to go on to publish *The Second Self*, exploring the relationship between children and computers), they explored the developmental psychology of children interacting with computers. It added fuel to the fire, and soon the pair was bursting with ideas, dreaming up schemes that would harness the power of the computer to playful learning.

The duo had complementary interests. Ocko preferred the physical world of objects, while Resnick felt most at home in the virtual world of software. When they cast about to develop their ideas into a toy, Legos, both very physical and yet virtually unlimited in their possibilities, seemed a natural foundation. A computerized Lego set could be used to endow the real-world experience of design with interactive depth, turning a static stack of bricks into a responsive object. Under Papert's guidance, the two began their work. Then something very fortuitous happened: Lego came for a visit.

In the autumn of 1984, the designers at Lego saw Papert interviewed on television and realized that Papert's ideas were remarkably similar to their own. Both stressed the importance of creative play as a fundamental learning experience for children. Lego had no agenda for the meeting; they just wanted to have a chat with the dean of educational computing. But when Papert informed Resnick and Ocko that the toy maker would be paying him a visit, they snapped into high gear, readying their prototype computer-controlled Legos for a demonstration.

Exceptionally successful at manufacturing toys without any active components, Lego had begun to entertain the possibility of extending their designs into more technological areas, not as an end in itself, but as an enhancement of the creative potential already present in the bricks. They had no firm plans and wanted to speak to Papert as part of a larger effort to fully grasp the capabilities of interactive aids to playful design. Lego certainly hadn't expected to see their future fully realized, but they were in for quite a surprise.

During the meeting, Resnick and Ocko presented a Lego system with an integrated system of motors, controlled by an Apple II running LOGO, Papert's child-friendly programming environment. This demonstration fairly blew Lego designers away, bringing the best of both worlds together in a way that diminished neither. In fact, these active bricks took the design possibilities to a new level. With strong encouragement from Lego, the two continued their work, refining their designs and adding mechanical components and software. In the summer of 1985, they dropped them in front of a group of children from a nearby school attending a two-week computer camp at the Media Lab.

The kids went nuts. They built a miniature amusement park with a variety of rides, all computer controlled, along with robots, houses with automatic doors, and a whole host of other inventive forms. Along the way, these children were learning about mechanics, physics, and programming, but they didn't even notice this; the learning was a means to an end, and play was the order of the day. When at the end of the camp, Resnick announced that these toys would follow them into the classroom (where Resnick would continue his research), the kids reacted with a mixture of disbelief and horror. This *couldn't* be schoolwork, they argued. It was fun. What does fun have to do with learning? This is a toy, it's fun to play with. Not school, nothing like it.

That's when Resnick understood how well he'd done his job. He'd wanted to focus on how children learn when they design things, believing design to be the best opportunity for a learning experience. Now he had proof positive that children *could* learn complex ideas quickly, without tears or pain, rote repetition or memorization, a transparent process of play leading into a richer understanding of the world. This research gave him the inspiration for his thesis, and in 1992, as he got his Ph.D., he joined the faculty of the Media Lab as a researcher in the "Lego Lab," in Papert's Epistemology and Learning Group.

Lego used the Media Lab's designs in a number of products. Meanwhile, Resnick and his team of graduate students kept plowing ahead. It took about four years to bring the first computer-controlled Technics kits to market. By 1988, Resnick was convinced they had further to go. These toys still required that the motors be physically attached to the computer—fine if you're making a static structure, such as a house, but very confining if you want to construct a robot or other autonomous object. If the Lego bricks could contain their own computing power, they could act independently—programmed via a computer and then sent into the world, untethered.

The programmable brick was the answer to this problem. Today the Media Lab focuses on ubiquitous computing, envisioning a world where computers are embedded everywhere in the environment, as they already are in the typical kitchen. Resnick and his team adapted this idea to their Legos long before the rest of the Media Lab had caught onto the idea of ubiquitous computing, creating a self-contained block with its own computer, sensor inputs, power outputs (to drive motors), infrared communications, and power source. The brick didn't need to be physically attached to a computer to communicate with one, so the computer could still be used as a programming environment. Once a program had been created, it could be downloaded to the brick, which would then run the program.

Resnick's team built a prototype and sent it out into the field—the museums, classrooms, and after-school centers eager for technological gizmos to spark the creative imaginations of children. Soon they found themselves surrounded by all sorts of autonomous toys: robots of different shapes, sizes and capabilities, all based upon the same brick. Even better, the children could use multiple bricks for a project. The infrared link that ran between brick and computer could also be used to pass messages—little bits of information—between bricks, so that the robots could act in concert, invisibly whispering their activities to one another, creating very complex behaviors from simple components.

That brick, christened the RCX, became the central element of Lego's Mindstorms. Although it took nearly a decade for Mindstorms to make its way from the Media Lab to the shelves of toy stores, Resnick and his team are gratified that their research has proven to be popular. "We believe there's a hunger for these things," he says, "but that's a hard sell, because many people believe kids won't respond to these kinds of toys. It's satisfying to know that there's a whole community growing up around them."

The success of Mindstorms reveals the genius of Resnick's design. The kit lives in a sweet spot between soldering together electrical components (too low level for any but the most hardcore hobbyist) and an inflexibly programmed toy that would soon bore most children. Mindstorms strides a middle path, giving its users just enough prepackaged technology to get them up and running without hampering their creativity. The child focuses on the creative endeavor while the RCX sweats the details. It's a potent combination.

Lego has shown their gratitude to the Media Lab by endowing the Lego Chair, now held by Resnick, and funding a $5 million "Lego Lab." Resnick's offices at the Media Lab are—to put it mildly—unconventional. A cavernous space with offices along its outer walls, the center of the lab features a pit of computers, toys, and a wall of Legos, neatly stacked into boxes of different sizes and colors. Graduate researchers sit in front of computer workstations, eagerly coding the next insanely great tool for educational computing or observing children as they play with toys. The whole space is decorated in bright primary colors, looking very much like a kindergarten for Ph.D.s.

This is no accident. Resnick believes that Friedrich Froebel, who created the first kindergarten in early nineteenth-century Germany, discovered something we tend to suppress as we grow into adulthood. Kindergartens present children with the opportunity to learn by playful exploration, design, and interaction. Historically, the enriched environment of kindergarten gets left behind as children progress into a more sophisticated understanding of the world, and the interactive possibilities of play are discarded in favor of books and other mediated learning materials. But children don't change as they grow older—they still acquire their knowledge in the same way. Resnick has become the advocate of lifelong kindergarten, where design, interactivity, and personal connection to the learning process become the integral features of education. Rather than reading about physics or math, Resnick wants children (and adults) to participate in an exploration that will feel much like play but will result in the acquisition of a broader base of knowledge—learning by doing.

If this seems obvious—and it is, to many parents—it can only be emphasized that we're not accustomed to educating our children according to these principles—at least, not once they've gone beyond kindergarten. Resnick is calling for a change in our educational mind-set. Through technologies like Mindstorms and the World Wide Web, children can become their own educators. If children seem reluctant to learn their classroom lessons, perhaps they'd be more apt to learn by playing. An educational environment designed to engage them through play could well unleash a creative explosion in childhood learning.

Resnick hasn't finished his work with Legos. Mindstorms, though selling well, represents just a midpoint in his inventive efforts. They're too big and expensive for a child to experiment with more than one at a time. If they could be shrunk down to the size of the standard Lego brick (the RCX is a monster compared with your average Lego) and manufactured even more cheaply, then children could use ten or twenty of them in a single project, creating truly novel designs from an array of active building components. Resnick has an obsession with complex systems, in which simple components work together to produce complex behaviors, much as Rodney Brooks's robots do. He wants to explore the educational possibilities of a whole "flock" of bricks working together on a single task.

Resnick already has his group working on another toy, the Cricket, a prototype for the next generation of Legos. Not very much bigger than the nine-volt battery used to power it, the Cricket has even greater capabilities than the RCX, but is one-eighth the size. And these Crickets are designed to chirp, able to communicate efficiently with one another, using infrared light to transmit complex messages that can be used to coordinate the activity of all Crickets participating in a given task. Using the Cricket, a child could create a flock of Lego robots, each communicating with its neighbors to work on a common task, much as ants or bees do in their colonies and hives. With these kinds of toys, Resnick speculates, we'll be able to help children learn about the complex systems that make up real life—things like traffic jams and stock markets and other phenomena that defy simple description—the kinds of experiences children will need to make their way in the twenty-first century.

We know that computers get smaller, faster, and cheaper with every passing day. The Cricket is probably just another midpoint in an increasingly active, increasingly tiny world of intelligent components. To understand just how small the Cricket will become in the next decade, we need to step back forty years and hear a few words from a man who said the most outrageous things—most of which happened to be true.

5 BUILDER'S BLOCKS

A LITTLE BIT OF GENIUS

Richard Feynman had the answer in the palm of his hand. Now he had to get the bureaucrats to listen to him. Waiting until the television cameras swung his way, he planned a bit of theater to get his point across. He'd bring a sobering close to one of the most horrifying events in the history of space exploration.

On January 28, 1986, the space shuttle *Challenger* proceeded through a normal launch sequence, got a few thousand feet into the air, and exploded, killing all seven members of the crew. The event, broadcast live into thousands of classrooms across the United States, amplified the tragedy, because NASA had chosen this mission to send Christa McAuliffe, the first teacher in space, into near-Earth orbit. This was to have been NASA's educational mission, bringing the wonder of space flight to millions of children. Instead, it set manned space exploration back for years; a generation who remembers the trauma of the accident will probably prefer to remain earth-bound in pursuit of knowledge.

President Reagan immediately convened a commission of experts to look into the causes of the explosion and to suggest appropriate remedies—such as changes in design or NASA policy—that might keep such a tragedy from repeating itself. Headed by Richard Nixon's secretary of state, William P. Rogers, the Rogers Commission brought together talents as diverse as

Neil Armstrong, the first man on the Moon, Robert Holz, editor in chief of *Aviation Week & Space Technology* magazine, and Alton Keel, director of the Office of Management and Budget—a mix of scientists and bureaucrats seemingly picked more for their marquee value than for their expertise.

Feynman, a physicist and Nobel laureate, had been reluctantly drafted from his laboratory at the California Institute of Technology in Pasadena, California. With a reputation as a scientific iconoclast, Feynman doubted that the commission could accomplish much of significance. He demanded (and received) carte blanche to interview NASA scientists and shuttle contractors and engineers, throwing himself into the work with an intensity that had made him famous forty-five years earlier, when, with a newly minted Princeton doctorate in physics, he joined the Manhattan Project at their ultra-top-secret Los Alamos facility. There Feynman outshone all of his peers—some of the best minds in the world of physics—in his ability to handle the complex mathematics of the atomic bomb.

After the war, Feynman joined the faculty at CalTech and applied his mathematical genius to the thorny problems of quantum physics, the subatomic world of particles and forces that make up the building blocks of the world of matter and energy. This work in quantum electrodynamics would win him the 1965 Nobel Prize in physics, but Feynman had broad interests in biology, computer science, and aeronautics; a restless intellect that impressed (and occasionally infuriated) the greatest minds of his day.

So Feynman made his tour, spending endless hours with NASA officials, asking them to detail their design decisions. The high-tech materials used on the space shuttle had been developed specifically for the kinds of stresses the shuttle would endure. He soon zoomed in on one particular part known as the O-ring, a rubbery gasket used to seal the shuttle's solid-fuel rocket boosters. These boosters, attached to the sides of the rocket, provided the extra thrust used to lift the shuttle into the upper atmosphere, where they were jettisoned back to Earth and retrieved for use on other shuttle missions.

These gaskets had a limited temperature range. Given that the shuttle was launched from the Kennedy Space Center in southern Florida, that point never worried NASA's engineers. But on the day of the *Challenger* launch, it was very cold on Cape Canaveral: thirty-six degrees at the base of the launch pad, and perhaps a few degrees cooler farther up the launch gantry—just where the O-rings sat. These rubberized parts didn't behave well in cold weather—they became brittle and wouldn't fill the gaps in the boosters, so they could easily leak fuel and cause an explosion.

During a public meeting of the Rogers Commission, Feynman, in full view of a constellation of television cameras, produced a small gasket made of the same material, dunked it into a Styrofoam cup of ice water for a few moments, then drew it out. The material, formerly quite pliable, had become rigid. A simple experiment that any NASA engineer could have performed in a few moments provided the answer to why the shuttle had exploded.

Several committee members were quietly enraged at Feynman's "grandstanding" and doubted his findings, but as more facts became known, it became clear that the genius's gift for dramatic flair hadn't lessened his firm grasp of the technical issues. Within a few weeks, Feyman's findings had been confirmed. This was the first time—and the last—that Feynman would come to the attention of the public. In less than a year, he'd succumb to the stomach cancer he'd been fighting off even as he worked with the Rogers Commission, leaving behind a growing reputation as one of the twentieth century's great minds, as captured in the autobiographical best-seller *Surely You Must be Joking, Mr. Feynman*, which portrayed him as enormously accomplished in the sciences yet profoundly human.

Consider a lecture Feynman delivered at CalTech in 1959 to a meeting of the American Physical Society, the prestigious college of the nation's physicists. "There's Plenty of Room at the Bottom: An Invitation to Enter a New Field of Physics" was, at the time, greeted with a mixture of laughter and amusement. Because he was well known for his scientific hijinks, most of the audience thought Feynman was putting them on, crafting an elaborate joke to entertain his peers in the physics community.

What was he suggesting that produced such mirth? Only that physicists had ignored one of the most obvious directions of technological development—the drive to make things smaller. These days, we take it as a given that computers and electronics grow increasingly small over time, but in the 1950s, when big was good and bigger was even better, this seemed counterintuitive. Transistors, only just coming into common use, were wired into computers that covered thousands of square feet. Atom smashers took up whole buildings. Nuclear reactors sprawled across acres. Big physics was good physics.

Five years after Feynman's talk, a semiconductor engineer by the name of Gordon Moore, who would later go on to cofound Intel, the microprocessor powerhouse, published a theorem that has since become known as Moore's law. It states that semiconductors (which compose both transistors and integrated circuits) improve at a fixed rate; every eighteen months they decrease in size by a factor of two and increase in speed by a factor of two. In the thirty-five years since Moore made this astonishing prediction, it has proven to be completely true: semiconductors are approximately *ten million times smaller and ten million times faster* than they were back in 1964.

Whether or not Feynman intuited Moore's law, he began his lecture with a bold statement:

> People tell me about miniaturization, and how far it has progressed today. They tell me about electric motors that are the size

> of the nail on your small finger. And there is a device on the market, they tell me, by which you can write the Lord's Prayer on the head of pin. But that's nothing; that's the most primitive, halting step in the direction I intend to discuss. It is a staggeringly small world that is below. In the year 2000, when they look back at this age, they will wonder why it was not until the year 1960 that anybody began seriously to move in this direction.

In an offhand way, Feynman was inviting his peers to join him in an investigation of the small. Physicists had been working with atoms and subatomic particles for over half a century—since the discovery of the electron—but had remained concerned with the theoretical qualities of these objects, mostly ignoring the practical implications of their research. Feynman proposed that it was possible to inscribe the twenty-four-volume *Encyclopaedia Britannica* on the head of a pin, using a technology similar to the electron microscope, just then opening the mysteries of the cell to biologists' eager eyes. If that could be done, Feynman mused, why not take all of the 24 *million* volumes in the Library of Congress, the British Museum, and the French Bibliothèque Nationale and place them onto a single sheet of plastic of about three square yards? Every person on Earth could carry the collected knowledge of humanity with them, in their pocket. A crazy idea, but physically possible:

> What I have demonstrated is that there *is* room—that you can decrease the size of things in a practical way. I now want to show you that there is *plenty* of room. I will not discuss how we are going to do it, but only what is possible in principle—in other words, what is possible according to the laws of physics. I am not inventing anti-gravity, which is possible only if the laws are not what we think; I am telling you what could be done if the laws *are* what we think; we are not doing it simply because we haven't yet gotten around to it.

What if we could take our machines and make them smaller? Feynman suggested building a machine, such as a metal lathe, at one-quarter normal scale, then using that to create another lathe at one-quarter its scale, and so forth, until, after a few generations, you'd be able to build a complex device—say, an automobile—just a few tenths of an inch in size:

> What would be the utility of such machines? Who knows? Of course, a small automobile would only be useful for the mites to drive around in, and I suppose our Christian interests don't go that far. However, we did note the possibility of the manufacture of small elements for computers in completely automatic factories, containing lathes and other machine tools at the very small level.

In just these few words, Feynman accurately predicted the at-that-time-invisible trend toward device miniaturization. The reason Moore's law works is because each incremental improvement in semiconductors is used to engender its succeeding generation. Results get fed back into the process; smaller parts are used to create ever-smaller parts.

And the smaller you get, the cheaper it becomes:

> Where am I going to put the million lathes that I am going to have? Why, there is nothing to it; the volume is much less than even one full-scale lathe. For instance, if I make a billion little lathes, each 1/4000 of the scale of a regular lathe, there are plenty of materials and space available because in the billion little ones there is less than 2 percent of the materials in one big lathe.
>
> It doesn't cost anything for materials, you see. So I want to build a billion tiny factories, models of each other, which are manufacturing simultaneously, drilling holes, stamping parts, and so on.

Feynman wasn't entirely starry-eyed; he noted that objects behave differently as they get smaller. Effects that wouldn't be noticed in the human-scale world become showstoppers in his miniaturized universe:

> All things do not simply scale down in proportion. There is the problem that materials stick together by molecular attractions. It would be like this: After you have made a part and you unscrew the nut from a bolt, it isn't going to fall down because the gravity isn't appreciable; it would even be hard to get it off the bolt. It would be like those old movies of a man with his hands full of molasses, trying to get rid of a glass of water. There will be several problems of this nature that we will have to be ready to design for.

After all of that crazy buildup, based entirely in the laws of physics—but sprinkled throughout with his unique brand of humor—Feynman got to the heart of his proposal:

> The principles of physics, as far as I can see, do not speak against the possibility of maneuvering things atom by atom. It is not an attempt to violate any laws; it is something, in principle, that can be done; but in practice, it has not been done because we are too big.

If we can scale down and get small, Feynman suggests, why not go all the way down to the bottom, to the atoms which make up the physical world, and learn to manipulate them directly?

At this point the assembled physicists were certain they were the victims of a classic Feynman put-on. Beautiful, elegant—and ridiculous. But they may have forgotten a famous line from the father of quantum physics, Niels Bohr, who crushed a promising graduate student's theories with the quip "Your ideas are crazy—but not crazy enough to be true!" Feynman's ideas were crazy. Crazy enough to be true genius.

GEEK SALON

At a certain point in every book—fiction or nonfiction—the author reveals himself to the reader. I've used the first person throughout this volume because I've been playing as I've been writing, but some of the events in this chronicle involved me directly.

The astute reader has probably already noted how frequently the Massachusetts Institute of Technology receives a mention in this text. There are at least two reasons for this: first, as a world-class institution, it attracts the kinds of researchers, such as Rodney Brooks and Mitch Resnick, who are worthy of note. The second reason is that I was a student at MIT, from the fall of 1980 to the spring of 1982.

I am no whiz kid who worked his way through the Institute at double time, finishing eight semesters' worth of work in four semesters. I was asked to take some time off, a euphemistically phrased request that really meant they didn't want me around. So I left and never went back—at least, not as a student. Instead, I started my work as a software engineer—which seems to be the default occupation of the Institute's legion of dropouts. (More than a quarter of the students who enter MIT as undergraduates don't finish their degrees.)

Whatever the benefits of an MIT education, most important to me was the exposure I received to some singular ideas and people. One of the residents on my floor at the Senior House, an MIT dormitory with the motto "Sport Death: Only Life Can Kill You," was working with Nicholas Negroponte on the still-evolving ArchMach. Another studied under computer science wizard and Disney Imagineer Danny Hillis as he developed the Connection Machine, a supercomputer built from sixty-four *thousand* microprocessors, all acting in concert. And some of my less well-grounded friends—the true geeks, living in a fantasy of a future world—hung on to every word spoken by a graduate student named K. Eric Drexler.

All of this proved tremendously influential to my wide-open mind; I may not have graduated from MIT, but the Institute left its mark on me. The *idea* of the Architecture Machine, which I never laid eyes on, drove me into my work in virtual reality ten years later. Danny Hillis inspired me to study complex systems and emergent phenomena. And Drexler—well, he shook my world to its foundations.

Eric Drexler has always had his mind on the stars. Coming to MIT as an undergraduate in 1973, Drexler studied in the interdisciplinary program, a unique custom-fit mélange of courses designed for students who could handle the workload in several different areas of academic concentration at the same time. The space sciences program fascinated him, and he studied biology, physics, chemistry, mathematics, and computer science—all core to the discipline. Drexler is one of those optimistic thinkers who believe humanity's future lies with the stars (perhaps he too had toy astronauts as a child) and holds that science represents the clearest path to the worlds of wonder beyond our own.

He stayed on to do his master's, and that's when he came upon Feynman's now-nearly-forgotten 1959 lecture. The idea of building things small has a great appeal to those in the field of space exploration; the smaller an object is, the less power it takes to lift it out of Earth's orbit and send it to the stars. If Feynman, a god of physics and another MIT graduate, had believed these tiny machines within the realm of physical possibility, then perhaps they could be built with the explicit purpose of exploring the other planets in the solar system—scaled-down versions of the Voyager 1 and Voyager 2 spacecraft then speeding their way toward Jupiter and beyond.

There is precedent for ultra-tiny machinery, and you're living proof of it. Literally. Each cell in your body has hundreds to millions of tiny organelles, minuscule wraps of proteins that perform specific mechanical functions. One of these, called the ribosome, acts very nearly like a computer. The ribosome translates the roughly 90,000 genes within each of your cells' nuclei into corresponding proteins. The wonder of it is that the ribosome is both so simple to understand and so completely improbable. Some three billion years ago, evolution produced a set of marvelous mutations, a way of coding the information about biological forms as a series of chemical bits (DNA), and, complementarily, created a "computer" (the ribosome) that read these bits to generate a nearly infinite variety of biological material.

In *There's Plenty of Room at the Bottom*, Feynman pointed to the success of biological systems as proof positive that microscopic machinery must exist—this before its existence had been even modestly explored. But for Feynman, no other answer seemed logical. Cells had to be built up out of these components, or else life itself would be impossible. Drexler, with the benefit of twenty years of progress in biology, could look at the scientific record and point to the reality of the molecular machine. He now began to wonder if human-engineered atomic constructions could improve on the organic mechanisms that had evolved within living systems. The molecular-scale design of active devices—*nanotechnology* was Drexler's name for these billionth-of-a-meter structures—could be the key to undreamt-of progress in space exploration.

Drexler badgered AI guru Marvin Minsky into becoming his thesis adviser. Minsky and Feynman had enjoyed a collegial relationship (artificial intelligence being one of the topics that excited Feynman's restless intellect), and Minsky knew of Feynman's theories about small-scale machinery. He'd also seen computers become increasingly powerful while shrinking dramatically in size. The handwriting was on the wall. Minsky, no stranger to scientific revolutions, signed up, spending the next decade guiding Drexler through his research, acting as his sounding board, and driving him toward an understanding of the awesome implications of his work.

When you assume, as Drexler did, that we can build anything, atom by atom, an entirely new universe suddenly appears. Things that are rarely found in nature, such as diamond, become manufacturable objects. Diamond is composed of carbon atoms arranged in a crystal of tetrahedrons extending indefinitely in every direction. A lump of coal, on the other hand, contains the same carbon, but in a random arrangement. This gives diamond its strength; crystals are hard to break apart, and diamond is the strongest crystal we know. But Drexler's hypothetical machines could piece apart the carbon atoms in coal and reassemble them, precisely, into a diamond lattice. Voilà! An inexhaustible supply of diamond, available in sheets (perfect to replace those delicate glass windowpanes) or in stones as large as you might like. Diamond is one of the most perfect building materials imaginable: strong, hard, and durable. Ideal for, say, a spacecraft.

Drexler hadn't wandered far from his roots, but even here, underneath his Tree of Knowledge, he saw plenty of good—and more than a bit of evil. An era of dramatic extension of human abilities lay before him, as soon as he got around to inventing the future. He came to understand how comprehensive the nanotechnology revolution could be, and this concerned him. Something this big would change human culture in unpredictable ways, much as the discovery of fire had quickened human civilization and helped to define our species a hundred thousand years ago. Convinced we needed to enter the Nanotech Era with our eyes open, Drexler began to talk about his ideas. And they attracted some followers. Including me.

One Saturday night in the winter of 1982, I joined two of my friends for a short trip from Senior House to a triple-decker flat in Cambridgeport, the dense swath of clapboard houses that fills the land along the Charles River between MIT and Harvard

University. I was being invited to a salon, a gathering of like-minded individuals for intellectual conversation about stimulating issues. Little did I know. I arrived at Eric Drexler's doorstep utterly unprepared for the revolution. I left his home a true believer.

Like any good evangelist, Eric had developed a concise, reasonable set of arguments about the feasibility and inevitability of nanotechnology. Look at the world of technology, he said, every year getting smaller and faster. All of the twentieth century has been a progressive refinement of manufacturing, from big to little. Within the next generation or so, we'll start to get down to the basic units of matter: atoms. When we get to atomic-scale assembly of devices, a whole universe of unimagined possibilities becomes realizable. What do you want? As long as it's physically possible, you'll be able to build it, as much of it as you desire. And, because we will have such fine control over the use of materials, it'll be cheap. Practically free. Nothing wasted, just the bare amount needed for any task.

I had received the Gospel according to Drexler.

At some point in the evening (my head grew so light I had to lie down for a while) I picked up a photocopied sheet laying in a small stack. "Ten Applications for Nanotechnology," it proclaimed, and proceeded with a list of items like ultra-strong materials (diamond); "meat machines," which take household garbage and scramble those atoms to produce a realistic facsimile of beef (or chicken, pork, etc.); immensely powerful computers smaller than a grain of sand . . .

It seemed impossible but utterly rational. Drexler is nothing if not methodical, reasoned, almost prosaic. Nearly the polar opposite of Feynman's intuitive intellectual omnivore, Drexler is the quintessential engineer working on the ultimate hack: nature. There was no way I could doubt him unless I was willing to throw my own reason onto the dustbin of history. I knew Eric was dead-on. It just made sense.

These parties—salons for geeks—aided Drexler in an important task: with so many eager and powerful minds at hand, he could talk through thc implications of nanotechnology with people who understood the laws of physics, chemistry, and biology, who knew computer languages and hardware design, individuals able to grasp the scope of the inevitable transformation confronting humanity. And make no mistake: it is inevitable. Even if the engineering of nanomolecular devices fails completely—and there's no indication that this will happen—we'll still be getting nanotechnology some time in the twenty-first century. But it will come from our own bodies.

There are two angles of approach to nanotechnology: molecular assembly, that is, putting together complex molecules one atom at a time, and molecular biology. That discipline, at the meeting point of biology and chemistry, has made enormous strides in the last twenty years, every day bringing new understandings of the chemical phenomena that create our vitality. Molecular biologists are at the threshold of understanding how the complex of biological machinery within our cells works as a unit to create the self-sustaining and self-organizing system we call life. There's a whole community of nanotechnologists who believe that nature is the best design guide for their own work, and molecular biologists are busily documenting the wheels and gears of this already existing nanotechnology. If nature can create a living cell from a complex of proteins, we should be able to build cellular-scale machinery from a similar foundation—or so the thinking goes.

If these arguments haven't proven persuasive, there's another, based upon Moore's law, which describes how quickly things get small. The semiconductors of the year 2000 are manufactured on a scale of .13 *microns*, or 130 nanometers (that's 130 billionths of a meter). Over the next twelve years, if Moore's law keeps pace (recently some scientists have begun to suggest that it's actually accelerating), we'll see 8 more doublings in speed, for a total increase of 256 times, and we'll also see structures 256 times smaller than those we manufacture today. We'll have broken through the nanometer boundary. An atom itself is about .2 nanometers across (sizes vary, based on the mass of the atom, its electrons, and so forth), so somewhere around the end of 2012 all of this miniaturization predicted by Moore's law *must* come to an abrupt stop; we will have reached the atomic level.

Nothing as sweeping as nanotechnology comes without its share of new worries, and in this case the devil—or rather, the devils—is in the details. These Drexler explored with his acolytes, all the while working his ideas (and theirs) into a text that would introduce nanotechnology to the world.

DOCTOR GOO

About a year after my first meeting with Eric Drexler, I got an excited call from the woman who had introduced me to him and who regularly attended his salons. She said, "Eric's written a book. Wanna take a peek?" An hour later, I returned home with a manila envelope containing a stack of loose manuscript pages—after receiving a stern warning to return the manuscript within twenty-four hours. I didn't need any prompting and stayed awake the whole night, working my way through the first draft of *Engines of Creation: The New Science of Nanotechnology*.

In 1986, Drexler published *Engines of Creation*, but the text I read in late 1982 or early 1983 bore only a passing resemblance to the book eventually published by Doubleday. The first draft, too mind-blowing for a mainstream audience, had to be significantly altered—softened—for Drexler to be taken seriously. So the published version begins with a sweeping statement of the importance of his research:

> The laws of nature leave plenty of room for progress, and the pressures of world competition are even now pushing us forward. For better or worse, the greatest technological breakthrough in history is still to come.

One thing about Drexler, he didn't understate anything. The first draft of *Engines of Creation* contained more speculative reality than most minds could bear. Certainly, as I zipped through the pages, I found revelations I'd never even begun to imagine, including the two essential elements of nanotechnology: the nanocomputer and the nanoassembler.

The nanocomputer, as the name implies, is a computer made up of atomic material. Today's large-scale computers, although composed of structures invisible to the naked eye, are still made up of hundreds of trillions of atoms, etched onto a silicon surface through a process similar to photography, where a negative exposure is used to create a positive image. These structures carry electrons—the basic units of current—through transistors connected together in blocks that perform logical functions, such as addition, subtraction, and comparison. The bits your computer manages are actually tiny units of charge—trapped electrons—which can be read as on or off. These two states describe the entirety of the world of the computer, which does nothing more than manipulate these bits, flipping them on and off as its instructions—also composed of bits—dictate. Every computer program, be it a word processor, a spreadsheet, or a web browser, is reducible to these two states, although it often takes tens of millions of bits to describe a complex program.

The contemporary design of computers isn't appropriate for a nanocomputer. Electrical charges on the scale used by today's computers, although quite insignificant, would destroy a nanocomputer, its atoms sensitive to even a single electron. Just one bit would fry its delicate structures, electrocuting it. To design the nanocomputer, Drexler threw his sights back to the mid-nineteenth century, long before electronics had been invented, to the prescient work of Charles Babbage.

Babbage, one of the inventive geniuses of the early Industrial Revolution, spent years developing something he called the Difference Engine, a terrifically complicated system of wheels and gears designed to perform arbitrary mathematical operations. The mechanical forerunner of today's computers, which are dramatically simpler by comparison, Babbage's engine could be programmed via a series of wheels, and with a crank, the machine would grind through an arithmetic operation, such as division or multiplication, to produce a result. His prototype demonstrated the soundness of his ideas, and Babbage received financing from the British government to produce a machine that could quickly calculate mathematical tables, a task that, until the mid-twentieth century, required an error-prone human calculator laboring for hours or weeks.

Babbage bit off more than he could chew. Rather than making incremental improvements to his engine, he wanted to create an engine so comprehensive that no further improvements would ever be needed. He established a friendship with Lady Ada Lovelace, the daughter of Romantic era poet Percy Shelley, and convinced her to write the first programs for the Difference Engine (yes, the first computer programmer was a woman). Although Babbage squandered a fortune on the Difference Engine, it never saw the light of day. Faced with dwindling funds and a distinct lack of enthusiasm from the prime minister (who really had no concept of the utility of such a machine, despite Babbage's repeated attempts to educate him), the project languished and finally died. A modern replica of the completed Difference Engine stands in Britain's National Museum of Science and Industry—and it works!

At the atomic level, mechanics, which we associate with a now nearly forgotten phase of industrialization, are once again the technology of importance. Rather than relying on the destructive flow of electrons, the gentle turn of a gear or spin of a wheel are manageable at the atomic level. In the natural world, atoms regularly form wheels in organic compounds such as benzene, a common industrial solvent—and Drexler has developed the design for atomic-scale gears. Put together correctly, these could be the foundational components of a nanocomputer, analogous to the transistor in present-day devices, but a thousand times smaller, and perhaps as much as a *million* times faster. (As structures get smaller, they increase their speed geometrically; a thousandfold decrease in size is equal to a thousand *times* a thousandfold improvement in performance.)

To store its bits, the nanocomputer would simply rotate the position of an atom. An up position might equal on, while a down position would be equivalent to off. This means that computer memory, currently composed of endless banks of transistors, each greedily holding the charge for a single bit of memory, would be reduced to a material in which each individual atom represented one bit of information. A grain of sand, translated to nanocomputer memory, could hold many more bits than all of the memory in all of the computers in the world.

How could such tiny structures be created? That's the job of the nanoassembler. Just as the ribosome, within our cells, translates our genes into proteins, the nanoassembler translates its own instructions, delivered via its onboard nanocomputer, into atomic-scale forms. An idealized nanoassembler would have a bin of different types of atoms—carbon, oxygen, hydrogen, nitrogen, and so forth—and would pluck an atom from the appropriate bin, attaching it to a growing chain, a molecule built one atom at a time.

Is such a thing possible? Biologists have learned that the enzymes which we use for digestion and which keep our brains running smoothly do exactly these kinds of finely tuned operations on a single atom of a single molecule. Of course, enzymes have no brains, operating blindly on whatever happens to be floating within a cell, but they do demonstrate a highly specific ability to create and modify complex molecules. If our enzymes can do it, a nanoassembler probably can, too. You might think of a nanoassembler as a type of ultra-miniature machine tool (to borrow a metaphor from Feynman), with a wide selection of arms, each suited for a particular task. Attaching a carbon atom? Okay, use this arm. Breaking two atoms apart? Use this one. And so on. The nanoassembler will need to be a multiheaded beast, widely adaptable to the volatile world of atomic chemistry.

There is a caveat: the best tool to create a nanocomputer is a nanoassembler, but a nanoassembler needs a nanocomputer to be anything more than a very simple piece of machinery. It's a classic chicken-and-egg problem; the nanoassembler really needs the nanocomputer to be broadly useful, a universal nanoassembler. But to construct a nanocomputer, you need a nanoassembler. One solution to this paradox would be to begin with a simple nanoassembler designed to build simple nanocomputers. Once nanocomputers begin to roll off its assembly line, the nanoassembler can be redesigned to incorporate the nanocomputer, and this more intelligent nanoassembler can be instructed to construct a more intelligent nanoassembler, and so on, in an evolving process of incremental improvement echoing our own biological history.

If nanotechnology has a Holy Grail, it's the universal nanoassembler. With it, any imagined nanotechnological device can be built. Once we have it, a universe of possibilities opens. Two such possibilities are particularly mind-blowing, and represent the heaven and hell of this new world.

In his lecture, Feynman made an offhand remark:

> A friend of mine suggests a very interesting possibility for relatively small machines. He says that, although it is a very wild idea, it would be interesting in surgery if you could swallow the surgeon. You put the mechanical surgeon inside the blood vessel and it goes into the heart and "looks" around. (Of course the information has to be fed out.) It finds out which valve is the faulty one and takes a little knife and slices it out. Other small machines might be permanently incorporated in the body to assist some inadequately-functioning organ.

I am holding Robert Freitas's *Nanomedicine* in my hands, published in 1999 and presenting an exhaustively comprehensive survey of the possibilities of medical nanotechnology. Although neither a doctor nor a molecular engineer (as nanotechnologists style themselves), Freitas has been studying both fields for years, and *Nanomedicine* exposes a depth of research that is simply mind-boggling. Beginning with the principles of nanotechnology (similar to what we've covered here, but rich in engineering details) Freitas builds a step-by-step case for the feasibility of nanomedical devices, mechanisms a millionth the size of a single cell, yet vastly more powerful.

One of the first targets of medical nanotechnology—a low-hanging fruit—is artificial blood. After the contamination scares of the 1980s, and with new diseases constantly appearing in the nation's blood supply, transfusions have become a serious issue in medicine. In a 1996 paper, Freitas describes his idea for something he calls respirocytes—named for respiration—nano-scale diamond spheres holding highly compressed oxygen. There's some basic atom-scale machinery inside the respirocyte, designed to release the oxygen on an as-needed basis, but it's only as complicated as a Scuba tank used by a deep-sea diver—just a billion times smaller.

The respirocyte doesn't appear to be difficult to manufacture—nowhere near as complicated as either a nanocomputer or a nanoassembler—and it may be the first commercially available form of nanomedicine. But we're just getting started. The last thirty years have seen an enormous improvement in medical scanning, beginning with computerized axial tomography (CAT) scans, which use high-energy X-rays to penetrate and image the body's tissues; progressing to magnetic resonance imaging (MRI), which literally magnetizes all of the hydrogen atoms in the body, then reads their "glow"; today, positron emission tomography (PET) scans give researchers detailed, real-time images of brain function, down to the level of tissues. All of these devices are comparatively blunt; it's difficult to make out specific groups of tissues, and the results are often very hard to read. For individuals with cancer, these scans are a matter of life and death—a misread scan can mean catastrophe, or a course of debilitating treatment where no treatment is actually needed.

Enter the medical nanobot. Also less sophisticated than a universal nanoassembler, the nanobot is equipped with a probe tip that can examine the molecular detail of the cell from the inside. Freitas believes such devices can be manufactured on the scale of approximately 200 nanometers on a side, considerably smaller than other cellular structures such as the ribosome. Once inside the cell (it would need to trick its way across the cell membrane, perhaps disguised as nutrition or a hormone), it could then read the contents of the cell, delicately recording data in its molecular memory. Once its scan was complete, the nanobot would exit the cell and transmit its findings (perhaps via ultrasonic vibrations, such as those used today on prenatal women) to a physician for analysis.

Of course, one nanobot looking at one cell won't tell you much of value. But a hundred million of them, still a vanishingly tiny volume, could spread out through the body, look for diseased tissues, cancerous cells, viruses, and so forth, and give a doctor a complete read of a patient's physical state in just a few hours. More efficient and less invasive than an MRI scan, from the patient's point of view, the medical nanobot could give us a complete picture of the human body and could provide early warning of any medical problems.

If problems should arise, another nanotechnology comes into play. The miniaturized surgeon fantasized by Feynman has become another of the goals of nanotechnology. Comprised of a nanocomputer, a scanner, and a nanoassembler, this nanosurgeon can locate damaged or diseased cells and *repair* them. Cancer, for example, is caused by a misreading of the genetic signals that control cellular reproduction. When these genes become damaged—through exposure to radiation, toxic chemicals, or inherited defects—the cell goes wild, reproducing indefinitely, creating tumors.

Broken genes could be fixed by the nanosurgeon, which would read the genetic code within the cell's nucleus, compare it with a clean copy scanned from healthy cells, then delete the offending codes, replacing them with the correct sequences. In some cases the damage might be too extensive to repair, so the nanosurgeon would simply inject the cell with a toxic potion, killing it.

If this seems a little far-fetched, remember that biologists and research physicians are using gene therapy today, trying to do exactly the same thing. A copy of the good gene, wrapped in a virus, is injected into the patient. The patient becomes infected with the virus—which means the virus inserts its DNA (the good gene) into the body's cells. Gene therapy uses the body's own machinery to repair damaged codes. If this process works (and it does, in a growing number of cases), genetic nanosurgery should also be possible.

When these technologies become available—and they will, over the next ten to fifteen years—medicine will be entirely transformed. Patients will go to the doctor for annual physicals during which they'll be injected with nanorobotic scanners. They'll sit in the waiting room while the probes make their scan. Then the patient will be read—the nanorobots will download their information to a computer in the physician's office. If problems are detected, the patient will be injected with specially programmed nanosurgeons, designed to locate diseased or incapacitated tissues and repair them. Patients might not notice any change in their health (it takes a while for cellular problems to avalanche into observable conditions), but they'll leave the patient in *perfect* health. As good as new. Indefinitely.

This last point was a major theme of *Engines of Creation*; if we can perfect medical nanotechnology sufficiently, Drexler argued, then the problems of disease and aging will become entirely manageable. Rather than fading into old age, we will remain vital—as long as we don't encounter the front end of a speeding bus or use a fork to get our Pop-Tarts out of the toaster while it's still plugged in. Nanotechnology doesn't promise immortality, but it does seem sure to grant us a vastly extended lifespan. (Statistically, a human being will have a fatal accident about every nine hundred years. Of course, when we have worked out the details of medical nanotechnology, we're likely to become a lot more cautious.)

In the first draft of *Engines of Creation*, Drexler went quite a bit further than this. Since a nanosurgeon can read your genetic code and correct it, why not use the genetic code to rebuild a human being who had passed from his earthly existence? We've all heard the urban legend about Walt Disney's corpse, preserved in liquid nitrogen, awaiting reanimation (pun intended). There are now several thousand cryonic pioneers who have had their bodies—or sometimes, just their heads—frozen in hopes of a day when medical technology will be able to revive them, cure their ills, and grant them another span in the world of the living. It does seem likely that a nanosurgeon could repair the damage caused by freezing, and could easily cure cancers, heart disease, and so forth. So perhaps the cryonic folks are onto something. Many of the nanotechnology pioneers (including Drexler) have signed up for cryonic suspension in case their demise occurs before medical nanotechnology has been perfected. Most of them believe in cheating death scientifically, and are willing to be a little ahead of the crowd.

The story does not end at death, however, and this was the most bizarre, most *unheimlich* part of Drexler's first text. Why stop at the cryonically frozen? Any body, as long as there's some brain left, should be a fit candidate for reanimation, particularly if the brain has been preserved (as it has, in the case of Einstein and more than a few others). It doesn't matter if the medium of preservation is toxic: a nanosurgeon can undo the damage (in many cases) and presto!—Einstein is back among us, or Tesla or Darwin. Anyone who has left us sufficient remains could be brought back from the dead.

You're probably thinking this is getting a bit creepy. I can imagine Drexler's editor at Doubleday clucking and saying, "This has *got* to go." Reanimation of the *very* dead is clearly implied in the published version of *Engines of Creation*, but never explicitly spelled out. Yet it looms as one of the more far-out but still entirely reasonable aspects of nanotechnology.

But there are simpler horrors awaiting us in the Nanotechnology Era.

If the nanoassembler takes atoms and patterns them into molecules, its inverse—the disassembler—can take any molecule and break it into its constituent atoms. A universal assembler would likely be able to perform either task adroitly. So consider this: as soon as a perfected universal nanoassembler is developed, its first instructions will be to build more copies of itself. (These devices can *reproduce*.) Furthermore, we could instruct that nanoassembler to use whatever material it finds in the world at large as raw components to make those copies. All of the copies produced by this nanoassembler could be given similar instructions. Pretty soon you'd have billions of nanoassemblers, and the number would be increasing exponentially, growing from a single instance to an unthinkably vast quantity in just a few hours. This is how nanotechnology works—machines building machines to build machines.

If such a nanoassembler got out of the lab, it would spell the end of us all.

Think of that nanoassembler as a virus; it lives to reproduce. The virus hijacks the machinery within the cell—your ribosomes—to make more copies of itself. The virus has natural limitations: if it kills you, it dies, too. But a nanoassembler is entirely self-contained. All it needs is a steady diet of atoms, and there are plenty of those in the world. The nanoassembler would eat whatever it found around it: a blade of grass, a lump of coal, or the neurons in your brain, converting them into other nanoassemblers. These would in turn repeat the cycle, eating more material, making more nanoassemblers, giving birth to another ravenous generation. And so on and so on and so on.

Estimates made by Drexler and others hold that if this ever happened, the entire Earth would be reduced to nothing but a swarming sea of nanoassemblers in as little as seventy-two hours. Lighter than air, and beyond the reach of gravity, they could travel the Earth invisibly, fanning out on the global jet stream, raining destruction—or rather, conversion—in their path. No place would be untouched. (In fact, they'd probably be light enough to escape the gravitational field of the Earth and head out to the other planets in our solar system.)

This problem—a real showstopper—was recognized early on in Drexler's research, and christened the "gray goo" scenario, reflecting the translation of the vital and variegated fauna of the planet into a singular and monolithic ecology of nanoassemblers. I've been frightened of gray goo for nearly twenty years. Even during the height of the cold war, no fears of nuclear war or nuclear winter ever matched my horror at the image of a nanoassembler infection, undoing 4 billion years of evolution's work in four days.

If, as a species, we made a collective decision to leave the nanotechnology genie inside the bottle, we might be able to avoid the gray goo problem. But the engines of history, constantly driving us to make things smaller and faster, to learn more about the basic molecular structures of life, seem to rule this out. And while Drexler and the nanotechnology community as a whole are a profoundly ethical lot (something we'll explore in the next chapter), we know all too well that science has both extended the horizon of humanity and brought it new terrors. Nanotechnology will provide the means to allow us to live nearly forever. If it doesn't kill us first.

What does this have to do with Lego robots? Everything.

6 THE MAGICAL WORLD

CONNECTED INTELLIGENCE

Welcome to mpesce's Playful World!
I'm a writer who has been playing with Legos for over thirty years.

If you could recommend any book that people must read, what would it be?
Understanding Media: The Extensions of Man, by Marshall McLuhan.

If you could put any message on a billboard for everyone to see, what would it be?
"Forgive yourself."

What's the most disgusting thing you've ever eaten?
Something made of chicken in Hong Kong.

When I unpacked my Mindstorms, I discovered a wallet sized plastic card with a secret ID number. The card encouraged me to visit the Mindstorms website at www.legomindstorms.com and log in as a registered user, gaining access to all of the special features that come with membership in a community of Lego enthusiasts. Okay, I thought, I'm game, and pointed my web browser to the site. In the Members section, I carefully filled out a form—using my secret password to gain access—and soon found myself creating a rudimentary web page, my personal domain in the Mindstorms kingdom. In addition to the usual name, rank, and serial number, the site asked me to provide a few words about myself, then asked three quirky questions (as given above) nearly guaranteed to shine some light on my inner workings. I have to admit they've given the matter some thought—I come across more honestly in these few words than in most anything else I've posted on the Web.

Lego developed Mindstorms with an explicit desire to create a self-sustaining community around the product. Not, as some might cynically suppose, as a marketing vehicle selling more toys to kids (though there is a complete product catalogue), but to encourage people to work, learn, and play together. Mindstorms, perhaps the most sophisticated toy ever produced, depends upon this community more than upon a sheaf of programming manuals or instructional aids. In fact, the Mindstorms kit has nearly no printed documentation, only a thirty-page *Constructopedia* detailing the physical design of some relatively simple robots. It's all about the computer through both the instructional videos I viewed on my PC as I assembled my Pathfinder One and the extensive Mindstorms website.

Once I have finished creating my home page in the Mindstorms universe, the site immediately invites me to contribute a design. The software tools provided with Mindstorms create programs that can be downloaded to the RCX, the brick with brains, but they can also be uploaded to the Mindstorms website. It works both ways. I am very interested in seeing what my more advanced peers in this community have created, because I still need to figure out how to re-create Rodney Brooks's Allen robot. I suspect it can be done, but I don't know how. My career in robotics has run through only the last two weeks, but much of what I need to know—about gears and wheels, motors and sensors—is contained in the freely shared designs of several thousand Mindstorms enthusiasts.

I start to click away, taking a close look at any of the designs I think might be relevant. First I come across a vending machine, which dispenses 1×2 bricks. That's pretty neat. Then a Lego sorter, which seems to be designed much like a coin sorter, programmed to take a messy heap of bricks and place them into separate piles. Next, I find a robot designed to assemble bricks into a particular form. Something begins to dawn on me. These systems, built with nothing more than Legos and a little intelligence, are the working demonstrations of the principles Drexler outlined in *Engines of Creation*. If Legos were atoms, I'd be looking at thousands of different nanomachines, all controlled by their onboard nanocomputer, much like the trusty and easy-to-use RCX if it were shrunk to infinitesimal proportions.

Dr. Ralph Merkle, a nanotechnology researcher, expressed a similar thought in a paper entitled *Nanotechnology and Medicine*:

> There is broad agreement (though not consensus) that we will at some point in the future be able to inexpensively fabricate essentially any structure that is consistent with chemical and physical law and specified in molecular detail. The most direct route to achieving this capability involves positioning and assembling individual atoms and molecules in a fashion conceptually similar to snapping together Lego blocks.

In other words, the experiential difference between playing with Legos and playing with atoms is nearly nonexistent. Both obey the same principles—up to a point—and playing with Legos can serve to prepare us for the world of atomic-scale devices. Which is why Mindstorms are so important.

For his Ph.D. thesis at MIT, Drexler designed an entire ecology of nanotech components, published in book form as *Nanosystems: Molecular Machinery, Manufacturing, and Computation*. Most of these designs exist only on paper, but some have begun to see the light of day. Researchers at the University of Michigan, which has a molecular design program, have created some simple gears, and at Yale, switches have been fabricated out of a few atoms of oxygen, nitrogen, carbon, and hydrogen. These breakthroughs, although admittedly leaving us far short of the universal assembler, point the way. Researchers, like children at play, are learning how to snap the atomic bricks together to create useful forms.

If we wanted to create a perfect playground for future molecular engineers, Lego Mindstorms would undoubtedly fit the bill. The set, flexibly designed, teaches children at play an important lesson: computer + bricks = device. As these children grow older, the bricks will grow smarter and smaller, blooming into nearly infinite possibilities. Apprentice nanotechnologists, confronted by a flexible universe as programmable and configurable as Lego blocks, will find nanotechnology the most natural thing in the world, easy as child's play.

Given the freedom and the opportunity to explore a new space full of possibilities, human beings instinctively scour the territory. First come the explorers, individuals like Drexler, who map the rough features of the land. They are followed by pioneers like Ralph Merkle (who recently joined nanotechnology start-up Zyvex) who stake claims and set up outposts. This paves the way for a flood of immigrants all dreaming of building a new life in a strange world.

Even before they have fully embarked on their fabulous journey, this particular community of immigrants has begun to recognize itself, gathered across the tenuous fibrils of the Internet, growing into a collective body of knowledge and craft welcoming all comers, provided they answer the call: *Design!* With each invention, the collective goal comes ever closer. The sharing of minds, across websites, e-mail, in the physical forms of robots and the abstractions of software, engenders an ever-broadening galaxy of forms, idea upon idea like so many bricks, creating a form still incomplete but nearing realization.

The connective intelligence of the Internet transforms the private activity of robot design into a collective process of discovery, paralleling developments in the nanotechnology community, which got onto the Web early and has come to rely upon it to knit far-flung research efforts into a coherent framework. Almost a decade before the Web had been invented, Drexler devoted an entire chapter of *Engines of Creation* to the importance of weblike technologies in the effort to create and manage nanotechnology. Given the inherent difficulty of fabricating nanotechnology, sharing knowledge made sense. And because of the dangers inherent to a realized nanotechnology, connective intelligence represents the *only* path to self-preservation.

In 1986, just as *Engines of Creation* was being published, Drexler founded the Foresight Institute, a nonprofit organization dedicated to the study of nanotechnology and its implications. From its base on the Web, the institute acts as a focal point for the nanotechnology community, creating the space for an emergent connective intelligence, one that could approach the radical transformation of the material world with forethought. In the afterword of the 1990 edition of *Engines*, he says:

> Some have mistakenly imagined that my aim is to promote nanotechnology; it is, instead, to promote understanding of nanotechnology *and its consequences*, which is another matter entirely. Nonetheless, I am now persuaded that the sooner we start serious development efforts, the longer we will have for public debate. Why? Because serious debate will start with those serious efforts, and the sooner we start, the poorer our technology base will be. An early start will thus mean slower progress and hence more time to consider the consequences.

A scientist asking his colleagues to *go slowly*? Drexler sounds ominously like nuclear physicist Leo Szilard, who refused to become involved in the Manhattan Project, fearing the product more than he loved the process. Although Drexler could rightly be called the Oppenheimer of this modern Manhattan Project—team leader, evangelist, and visionary—he has driven the collective project of a nanotechnological future with singular humanity, never forgetting that the ends do not always justify the means.

In the early morning of July 16, 1945 as Oppenheimer witnessed the explosion of the first atomic bomb, his mind locked onto a passage from the Bhagavad Gita, the Indian epic of conflict and indecision: "I am become Shiva, the destroyer of worlds."

Drexler hopes never to have cause to utter these words.

SPACE CASE

Somewhere in cyberspace, visiting the Nanocomputer Dream Team Home Page:

> Through the power of the Internet, talent from all over the World in every scientific field, amateur and professional, will rise together to create the World's first Nano-Meter SuperComputer.
>
> (Click.)
>
> Welcome to the coordination centre for the future of computing! The NanoComputer Dream Team is an international group of scientists, academics, researchers, and engineers—all working together on various projects to build the world's first nanometer-scale supercomputer.
>
> **You Can Help!**
> We want you to join the Team. We're looking for volunteers to join our project teams—it doesn't matter if you're only willing to

donate a bit of spare CPU time, or you want to get heavily involved. You'll join a team of others in our project team intranet, helping develop your team's component.

Project Underway
Our distributed computing team is about to get underway with its modeling program. Here's a screen shot of our prototype screen saver that will help share CPU cycles. To get involved, join a Team.

These guys (and I assume it's mostly guys) really have their act together. They've figured out that the Internet is a multiheaded beast, and they're throwing everything they have out there, hoping some of it will stick. They might be hopeless geek romantics, Trekkies dreaming of a transcendent destiny, but they talk the talk. I snoop around a bit and find out these pages were posted in 1996. An eternity ago. Perhaps their efforts were stillborn. But from these pages I stumble onto the NanoRing, a web ring of cross-linked web pages, each pointing to its neighbors in cyberspace like so many beads on a string.

Nearly all of the NanoRing sites have been created by amateur nanotechnology hobbyists, intelligent laypeople to whom the thornier problems of nanotechnology won't yield themselves. Folks like me, I'd imagine. A few more clicks and I find myself in the Tools section of the Nanotechnology Industries website, scrolling through a fairly lengthy list of software, from million-dollar molecular design programs (used by pharmaceutical companies to design new drugs) to NanoCAD, a free program for my web browser that'll allow me to do some molecular engineering. Click.

My web browser has just become a simulator for nanotechnology. All of a sudden I *can* design nanomachines! Okay, I read the too-brief instructions and begin to mouse around, creating six carbon atoms, dark gray circles on a light gray background. Then

I bond the atoms together, dragging my mouse from one circle to the next, until all are connected. I do it again for good measure. Now (if my knowledge of organic chemistry can be trusted) I've created benzene, the simplest of all the atomic ring structures, and the theoretical foundation for a nanomachine gear. Actually, my molecule looks rather squashed—benzene forms a perfect hexagon—but it's not bad for a first try.

At some point in the not-too-distant future I'll be able to hit the Print button—as if I were sending one of these chapters to my laser printer—and my simulated atoms, carefully chosen, will become the stuff of reality. This nearly magical molecular manufacturing is closer than you might think—something we'll discuss a bit further on. Even with my new capabilities, I don't have the know-how in chemistry or mechanics to design anything more than very basic components. A contraption as advanced as Drexler's atomic-scale Difference Engine is completely beyond me; even if I had the detailed plans of Charles Babbage's mechanical computer, the Difference Engine, I wouldn't know how to translate them into the appropriate molecular components. That's a job for rocket scientists. Like the ones at NASA.

NASA's Ames Research Center in Mountain View, California, has a history of cutting-edge scientific discovery. The first modern virtual reality systems, created at Ames in the early 1980s, defined the field. Today Ames's scientists are working on another far-out problem: the design of the nanocomputer. The Computational Molecular Nanotechnology team, formed in early 1996, has the stated goal of developing computers that can execute a *billion billion* instructions per second (in comparison, most home computers run at around half a billion instructions per second), using memory that occupies a *million billionth* of a centimeter per bit.

These are pretty esoteric targets, even for NASA. But Drexler's original vision of light, low-cost, and highly flexible systems for space travel and exploration has been accepted by the astronautics community as an important component (should it prove possible) in the future of the space program. Why? Consider the Mars Polar Explorer, which disappeared without a trace in early December 1999 as it approached its landing on Mars's southern polar ice cap. The craft, which cost a bargain-basement $165 million to design and launch toward the red planet, was still big enough to be destroyed by a thousand natural or man-made accidents. The landing sequence actually featured a controlled crash onto the planet's surface, using a system similar to the air bags in automobiles to absorb the stresses of impact. (Recent findings seem to indicate that the Polar Explorer's crash did it in.)

The sophisticated electronics and probes contained on the Polar Explorer could be duplicated at the atomic level, once the technology has been sufficiently perfected, leading to a spacecraft of similar capabilities, but only a few millionths of a meter in dimension, roughly the same size as a human cell. Within this volume, an unimaginably powerful nanocomputer with a comprehensive memory, chemical analysis system, and communications module could reproduce all of the Polar Explorer's features and still be lighter than air. And if you can build one to send to Mars, why not send *fifty million* of them—all in a volume weighing less than a pound! Even if as many as half of them malfunction along their seventy-million-mile voyage, there's still twenty-five million left to fan out, gather data, conduct analyses, and map the unknown.

Forty years ago, as a consequence of President John Kennedy's announced goal of sending a man to the Moon, and returning him safely, NASA inspired an earlier generation of miniaturization: the integrated circuit. The complex systems on the Apollo launch vehicle couldn't be made out of discrete components (bulky, power-hungry transistors) and still be light enough to reach the Moon. So NASA went to companies like Texas Instruments and Fairchild Semiconductor, and asked them to weld these components into a single device. Thus was the integrated circuit born. Although few integrated circuits made it into Apollo, the integrated circuit became the foundation of the Information Age, powering computers, cell phones, and CD players, enabling the Internet and the Web to span the planet. NASA set the stage for the age of miniaturization, and seems prepared to take it the rest of the way, down to our atoms.

This will not be easy work. Ralph Merkle, who acts as an adviser to the team, predicts that even if everything goes according to plan, if they encounter no insurmountable roadblocks on the path to the fine manipulation of molecules, he expects the first nanocomputer components to be available no sooner than 2011. It will take time to build these components into generalized nanocomputers, but at that point, they could become the vital, intelligent hearts of universal nanoassemblers. And while nanotechnology requires less of a scientific leap of faith than does the creation of artificial intelligence, it could fail as utterly as old AI did back in the 1980s. The only way to know for sure is to try.

At the turn of the twentieth century, most scientific experts claimed that heavier-than-air flight was impossible, but two amateurs, bicycle makers from Ohio, built an airplane. Positive proof against negative theories. This time the amateurs, looking on from the sidelines, cheer the experts on as they work their way down into the heart of matter. If nanotechnology proves a plausible dream, these amateurs will become inventive talents in their own right—pent up with twenty years' worth of imagination—using the engines of creation to fabricate a new world.

PRIVATE LANGUAGES, PART II

NASA, for all their good intentions, might also deliver the technology of infinite destruction, a toxic shock that could sweep the Earth clean. Not even the cockroaches, those radiation-hardened and mutation-resistant bottom feeders of the insect kingdom, would survive. Drexler scared himself so well with his own apocalyptic fantasies of replicating nanoassemblers—or replicators—drowning the Earth in an orgy of reproduction that he later backed away from his frightening hypothesis:

> Today I would emphasize that there is little incentive to build a replicator even resembling one that can survive in nature. Consider cars: to work, they require gasoline, oil, brake fluid, and so forth. No mere accident could enable a car to forage in the wild and refuel from tree-sap: this would demand engineering genius and hard work. It would be likewise with simple replicators designed to work in vats of assembler fluid, making non-replicating products for use outside. Replicators built in accord with simple regulations would be unlike anything that could run wild. The problem—and it is enormous—is not one of accidents, but of abuse.

Human beings require trace amounts of certain organic materials, without which our biological processes come to a screeching halt; they're known as vitamins and are essential to our continuing health. Drexler suggests that replicators be designed with "vitamin deficiencies" that can only be accommodated in very specific environments, laboratories similar to the ones we use for dealing with lethal viruses, like Ebola. This part of a future nanotechnology industry will probably look a lot like the biotechnology industry, which runs under federal guidelines to prevent contamination of the public sphere. This may suffice to keep us safe—but not indefinitely.

There's little incentive to build nuclear weapons—only two have ever been used for anything other than experimental purposes—but governments believe that the threat of mass destruction satisfies policy goals and frightens potential enemies. Although nanotechnology is today's rocket science, after the basic breakthroughs of the nanocomputer and nanoassembler, these techniques will become off-the-shelf components in a larger library of possibilities. Drexler may design a replicator with an eye to safety; a government facing terrible enemies might not. Like the nuclear genie, which reconfigured the nation-state system into the strategic blocks of the cold war, the dark arts of nanotechnology will force us to come to terms with a frightening truth: the possibilities for planetary extinction will become increasingly available to a wider range of humanity.

Great social ills have historically been treated by the physicians of culture through the balm of education. Meet Eric Drexler, toy designer:

> Picture a computer accessory the size of your thumb, with a state-of-the-art plug on its bottom. Its surface looks like boring gray plastic, imprinted with a serial number, yet this sealed assembler lab is an assembler-built object that contains many things. Inside, sitting above the plug, is a large nanoelectronic computer running advanced molecular-simulation software (based on the software developed during assembler development). With the assembler lab plugged in and turned on, your assembler-built home computer displays a three-dimensional picture of whatever the lab computer is simulating, representing atoms as colored spheres. With a joystick, you can direct the simulated assembler arm to build things.

Sounds a lot like NanoCAD with a working Print button, or Mindstorms gone very, very small. At MIT's Media Lab, you could say Mitch Resnick and his team are designing the future interface for Drexler's toy—different trajectories on convergent paths. But down here, in the domain of the very small, the mix of the real and the virtual becomes confused; is it simulation or reality? This is the new playground of the real.

Our millennial child, now hovering on the first fringes of adolescence, receives another gift on another Christmas morning, fit for a child coming of age in an era of extraordinary responsibility. This is the way the world works, her parents say; go and explore. She finds herself with a walnut-sized shell of boring gray plastic, which she places beside the computer in her study. An ultra-fast link connects the two instantly, and she beholds the universe contained within that plastic nut, raw piles of atoms spread like so many grains of sand on the beach. It seems impossible to her that complexity could come from this boundless simplicity. How can she even begin?

Instinctively she reaches out for her web browser, typing in the address printed atop the toy. *Welcome to the Guild of Material Adventurers,* the site advises. *Log in, apprentice, and begin your journey.*

This is carbon, the basic building block of the living world. She's heard this in school already, knows that plants and animals alike are composed of the substance, but in endless variations. *If you place carbon atoms together like this, you can make a crystal known as diamond.* Soon she's created a fleck of diamond just a hundred millionth of a meter wide, deep inside the plastic shell. She can see it on her computer screen. *Carbon atoms can also be placed into a ring, known as benzene.* She's quicker now, and does this in just a few moments. *Benzene rings can be used as gears.*

Now she walks through a lesson in mechanics, connecting benzene rings together to create a nano-scale clockwork, learning about gear ratios and torque. *You can use a carbon spring to drive this clock.* The spring is easy to create, just a chain of carbon atoms bound in a tube a few nanometers wide. *Twist the spring to add energy to it.* Soon she's got a working clock, nearly identical in design to those manufactured four hundred years ago. *Congratulations, apprentice.*

For her reward, she is granted access to the grimoire of the guild, an encyclopedia of spells and incantations, the result of years of work by millions of minds, each adding their own distinctive additions to the growing library of forms. Nearly everything she encounters looks complex beyond comprehension, but she can recognize some of her clockwork designs incorporated as elements within these works. One of the entries seems easy enough, just a more sophisticated combination of the gears she's already worked with. *This is an adder. It takes two numbers and adds them together.*

From the adder, she moves to a multiplier, a divider, and another tiny machine which compares two numbers. Now she learns the logic of on and off, and from this creates a basic calculator that can run a small program. *Congratulations. This calculator can be used to create a computer.*

Mastering these building blocks, arranging them carefully, connecting one to another like so many Tinkertoys, spoke to hub, wheel to chain, gear to motor, she creates a fully capable computer, which she programs using another tool provided by the guild. *This program will allow your computer to build other computers.* She sends the instructions to her creation, like the sorcerer's apprentice conjuring an army of genies to do her bidding, commands them to multiply, divide, expand until every space has been filled, her little toy overwhelmed with the ever-increasing generations of her creation.

Frightened that her toy has malfunctioned—worried she's broken it somehow—she presses the Reset button, and chaos is restored from the drowning sea of order. *This is what happens when you build machines that make copies of themselves.* She has witnessed the end of the world now, and nods in understanding. *Much is possible, not all of it safe.*

As a journeyman, she casts about for a project that will mark her own addition to the grimoire. She studies the growth of plants, the subtle mathematics of form that spell beauty from a simple equation endlessly repeated. In the branching of trees and unfolding of petals, math and biology collaborate in design, so she runs a hundred, then a thousand, then a million combinations, each subtle variations on a theme, weeding out most, saving a few, breeding these with each other. The variety soon exceeds her own expectations.

A perfect rose, five hundred millionths of a meter tall, becomes her gift to another generation of apprentices.

III

PRESENCE

7 UNDER THE SUN

RULE BRITANNICA

I have to confess: I've been cheating on my research. Okay, not quite cheating. But rather than sitting in a library, pouring over stacks of books and periodicals as I compose the text of this book, I quite frequently open up my web browser and point it to www.britannica.com. For example, I might type in the words *World Wide Web*, hit the Find button, and discover the following:

> **World Wide Web**
>
> (www), byname THE WEB, the leading information retrieval service of the Internet (the worldwide computer network). The Web gives users access to a vast array of documents that are connected to each other by means of hypertext or hypermedia links—*i.e.*, hyperlinks, electronic connections that link related pieces of information in order to allow a user easy access to them. Hypertext allows the user to select a word from text and thereby access other documents that contain additional information pertaining to that word; hypermedia documents feature links to images, sounds, animations, and movies . . .

When I need historical facts about the personages discussed in this book, people like Jean Piaget, Norbert Weiner, or Charles Babbage, I turn to *Britannica*. With just a few keystrokes, I'm deep into the well-researched and carefully cross-referenced articles that describe these men, their work, and its significance. First published a quarter of a millennium ago, the *Encyclopædia Britannica* remains the definitive reference work of the English language, the launching point of Richard Feynman's lecture "There's Plenty of Room at the Bottom," and the standard-bearer for intellectual completeness. Found in every library in the English-speaking world, where it generally occupies an entire bookcase, *Britannica* has served as the foundation of countless term papers, research reports, and surveys, a base of collected knowledge that, while considerably smaller than the total knowledge base of humankind, still comes to over a billion characters of printed text.

Then the Web came along.

In the second half of the 1990s, reference texts such as *Britannica* took a backseat to the instant knowledge available through online resources: Web pages appeared on the most curious of subjects, some factual but incomplete, others in depth but too technical, and many just plain wrong. It took some time, but the editors at *Britannica* decided to take the plunge. On October 19, 1999, www.britannica.com launched on the World Wide Web. And immediately crashed.

It seems that so many people were hungry for the solid facts in *Britannica*'s virtual pages that they simply overloaded the web servers. Something like fifteen million hits—individual queries to *Britannica*'s knowledge base—were recorded by the system before it overloaded and expired, leaving *Britannica* a bit red-faced with embarrassment. Clearly they had underestimated the online demand for such an important information resource. The technical wizards at *Encyclopædia Britannica* dusted themselves off, ordered some more powerful computer equipment and a "fatter" connection to the Internet capable of handling the expected traffic load, and brought the site back online in mid-November. Just in time for me to put it to good use.

How did I learn about this brief disaster? I got that information off the Web as well, from the knowledge base of technology news stories at www.wired.com—the online arm of *Wired* magazine. Nearly all of the facts I present in this text, the stories I use to make my case, have come from my research on the Web. This isn't a bad thing, not at all: it means that my facts are up-to-date, and if I play close attention to my sources, quite reliable.

At this point, I can't imagine doing my work any other way. Certainly there are some things I can't get from the Web, some islands of information so tightly controlled that they can't be reached by my browser. But it seems unlikely that this information becomes more valuable if fewer people have access to it. In the twenty-first century, the value of information seems to be directly proportional to how easy it is to access. Fast facts are worth more than slow ones.

The folks at *Britannica* seem to understand this as well, for their website includes not only the entire encyclopedia but also the *Merriam-Webster Collegiate Dictionary*, with 100,000 English words, and a comprehensive search engine: when I type in "World Wide Web," I get the entry from *Britannica*, another from the dictionary, and a page full of links to other web resources *Britannica* has deemed relevant. I needn't stop with *Britannica*; I can start there, then fan out to study the entirety of the Web, beginning with a simple search. (I did just this when studying the fabulous career of Jean Piaget.)

Somehow all of this seems very ho-hum, just another technological marvel in an era that has seen an ever-increasing supply of them, a diamond thrown onto a growing pile of jewels. But my way of working *is* different; it is unlikely that I'll ever bury myself in reference books again, because the information I need is everywhere around me. As long as I have a computer and a phone jack, I can be nearly as well equipped as someone in the Library of Congress. And it will only get better.

Most of us haven't adjusted to this new atmosphere, where knowledge has become instantly and pervasively available, except in very specific situations, such as those fabled day traders who scan the Web for financial news, corporate reports, and the dirty rumors that can raise a stock's price to stratospheric levels or send it into the basement. As a whole, we work much the same way we used to, reflecting the ways we were taught to acquire knowledge as schoolchildren.

Children born today will never know a world without the Web. It is the air they breathe. And that simple fact makes all the difference in the world.

MISSIONARY POSITION

In 1908, British theologian and hermeneutist Dr. Frank Charles Thompson published his *Thompson Chain Reference Bible*, the result of twenty years of dedicated study of the Christian scriptures. In addition to the standard King James text of Old and New Testaments, his Bible, along its outer margins, contained detailed links directing the reader to other Bible passages. For example, passages about humility, which are legion in the Bible, might lead the reader through various Old Testament prophets into the Gospels and finally to the Epistles, a chain of connections that Dr. Thompson hoped would explicitly show the authority and continuity of the Bible.

Nearly a century later, Thompson's Bible has become a standard-issue item for preachers around the world; they research their own sermons—often based on the themes explored by Thompson—by finding the appropriate entry in the Thompson Bible and tracing the connections through the text as they build their arguments. Although a Christian might read the Bible continually throughout the span of a religious life, the text is too large to know exactly what's where. Thompson made this task much easier, providing crib notes for the evangelically inclined, clearing paths through the thicket of the Word.

The *Thompson Chain Reference Bible*, entirely self-contained, points only to itself. All of the links reference other passages in the Bible, and while this is fairly comprehensive, it remains an island unto itself. Other versions of the Bible place different translations of the text side by side, for comparison and meditation, often with the original Hebrew or Greek passages of scripture. With these works, it becomes possible to understand the minds of the translators, their assumptions and mistakes, to capture the meaning hidden behind the words. Robert Graves, the British author most famous for his historical novel *I, Claudius,* wrote extensive commentaries on the Bible based upon his own readings of the original manuscripts and other accounts within the same historical period. The Jewish historian Josephus, for example, gives us the first reports (93 A.D.) of the Christian cult then springing up in Jerusalem and Rome; this Graves used in his own novelistic retelling of the Passion, *King Jesus.* Graves's story reads very little like the Gospels, but as he points out in his introduction, has as much historical authenticity as the New Testament. Graves used the same threads, but drew them into a different weave.

Our thought processes work along much the same lines. Two people might reach the same conclusion through very different trains of thought. Our thinking is associative: one thing reminds you of another thing, which connects to another, and so on. These chains of reason, like beads on a string, create our understanding. That's a very personal process (the thoughts inside my head probably bear little similarity to yours), but we can lead others down our trains of thought, jumping from one step to another, to arrive at a mutual understanding.

Books visibly illustrate this process; while reading *The Playful World* you're exploring how my mind works, following my arguments, gaining insight into what makes me tick, a peculiar process we've been able to share with one another since the invention of writing some 5,500 years ago. Most books are composed of such implicit links, referring back to the author's experiences or readings.

This idea obsessed Theodor Holm Nelson. In the late 1950s, this son of movie star Celeste Holm attended Swarthmore College. Nelson wrote a movie, acted in plays, went to classes (sometimes), but mostly dreamed up schemas, vast structures of logical relationships that could express, in rich detail, the flavor of human existence. Language seemed incomplete to Nelson, so he dedicated himself to the creation of normatics, a comprehensive system that would allow people to "say complicated things more clearly."

Clearly a complicated thinker himself, Nelson filled boxes with 4×6 index cards, each of which contained a single thought or idea or rumination that had crossed his mind. As Nelson tried to index and organize these cards, he found that many of them belonged in multiple categories. He developed a system of markings on the upper left-hand corner of the cards; these markings let him know which subjects might be covered on a particular card. Still, there was no way to sort these cards; any ordering would be incomplete. Nelson began to imagine a new medium that could accurately depict the implicit connections between cards. This medium would illustrate the connections between his cards and his thoughts.

The idea grew on him. If he extended it outward beyond his own notes, he could imagine linking his own writings to another person's, from that person to another, and so on, until the entire thought of humanity became a cohesive whole that could be traveled in depth. He could dive into his own ideas, then swim the sea of connections, leaping beyond the boundaries of paper into a continuous field of knowledge.

An iconoclast and firebrand even in his college days, Nelson railed against the idea that people could not or would not think for themselves when pondering the important issues of the day. During the 1960s, as Nelson worked through his ideas, matters of significance could be found in every coffeehouse and at every demonstration. Free speech became synonymous with freedom itself, and Nelson imagined himself as the champion of a new kind of liberty—liberty of thought. Nelson wrote of this in his autobiography, *World Enough:*

> As those weeks of Fall 1960 passed, I tried to tell everyone my growing vision. It was obvious, unified and cosmic.
>
> We would not be reading from paper any more. The computer screen, with its instant access, would make paper distribution of text absurd.
>
> Best of all, no longer would we be stuck with linear text, but we could create whole new gardens of interconnected text and graphics for the user to explore! (It would be several years before I would choose the word "hypertext" for this vision.) Users would sit holding a light-pen to the screen, making choices, browsing, exploring, making decisions.
>
> This would free education from the tyranny of teachers! It would free authors and artists from the tyranny of publishers! A new world of art! knowledge! populism! freedom! The very *opposite* of what everyone thought computers were about!

It's difficult, some twenty years into the personal computer revolution, to remember that computers were once thought of as monolithic, sterile entities, programmed by punched cards stamped DO NOT FOLD, SPINDLE OR MUTILATE, and tinged with a vague evil, almost as if they were thinking up ways to dehumanize us. Computer screens simply didn't exist outside of a few laboratories. Nelson had seen a photograph of a computer monitor in a magazine, put the pieces together, and got a glimpse thirty years into the future. Computing would be personal (everyone would have their own screen) and we'd use our individual computers to connect all human thought.

Ted Nelson had a mission. Five years later, he gave it a name: *hypertext.*

When you meet Ted Nelson, you're immediately struck by his boundless energy, an enthusiasm that seems to seep from every hair on his head. He appears to generate several new ideas a minute, concepts that pop out with such abandon he hangs a notepad around his neck to jot them all down. His excitement is infectious, and—particularly to young people—absolutely captivating. Nelson had trouble seducing the white-jacketed computer technologists with his visions, but he had rather more success with his peers, including one named Andy Van Dam.

Nelson had an Achilles' heel—he didn't have access to a computer. In the 1960s, they cost hundreds of thousands of dollars and filled rooms. While fully able to envision the systems of the future, he lacked the ability to translate these ideas into reality. So he made a point of befriending people who did have access to computers, beginning with Van Dam. The two grew close, and gradually Nelson persuaded Van Dam (who was already busy developing the fundamentals of computer graphics) to create a prototype hypertext system, something that would show the world what he was talking about. Nelson knew that a single demonstration would be proof enough.

Nelson's design and van Dam's code came together into a simple application that demonstrated the inherent connectivity of ideas, a vision that had consumed Nelson for most of an entire decade. They wrote a paper about their work and got it published in a scientific journal. Meanwhile, a friend informed Nelson that a scientist in California shared his big ideas.

Alongside Ted Nelson, another visionary had been puzzling through a slightly different question: how can we use computers to augment the human intellect? Douglas Engelbart had been asking himself this question since the late 1950s, and, as a researcher at the Stanford Research Institute (SRI), a think tank spun out of Stanford University, he began to tackle the problem. Engelbart had already established himself as a first-class researcher, and had worked up so many patents in his first years at SRI that his managers gave him carte blanche to pursue his own passions.

Like Nelson, Engelbart is a man of deep vision, and the world he saw at the end of the 1950s little resembled the quiet years some fantasize about. Engelbart foresaw future shock, a world where change could be the norm rather than exception, and he wondered how human beings would cope with the accelerating pace of change. He imagined we'd need to make quantum leaps in our own understanding just to stay in place. But how?

Engelbart's answer: People need to be smarter, and they need to apply their smarts to getting smarter, constantly *augmenting their intellect*, using intelligence in a leapfrog game to gain ever greater intelligence. As Piaget did half a century before, Engelbart started with an examination of the innate capabilities of human beings, looking at the natural gifts of intelligence, then studied how they grow into knowledge, skills, techniques, and cultures. These skills had grown slowly, building up over the millennia of human civilization, but Engelbart suspected they could be bootstrapped into enormous new capabilities with digital technologies. If he could engineer the proper set of tools and skills, Engelbart thought he might be able to create an intelligence augmentation process that could evolve with humanity, constantly creating smarter tools and smarter people, who would create even smarter tools for even smarter people.

The centerpiece of this process—the most important tool—became known as the oNLine System (NLS), a computer program through which groups could pool and share their knowledge. NLS allowed researchers working at computer terminals—just coming into common use—to create collaborative documents, featuring links between documents (like Nelson's hypertext), videoconferencing, and directories of online resources. In order to facilitate interaction with the NLS, Engelbart invented a tool that could be used to navigate NLS, a palm-sized device that rolled about on the tabletop and could be activated by touching buttons located on its surface. In order to make the NLS easy to use, Engelbart invented that most common of desktop appliances, the mouse!

Many of the features we think of as common in desktop computers—such as the mouse, windows, electronic mail—were pioneered by Engelbart, all answers to the question of intelligence augmentation. Engelbart may be the final proof of the value of NLS; he used it to further his own work, multiplying his own intelligence to produce some of the most remarkable innovations ever made in computing. When Nelson met Engelbart in 1967, the two found a camaraderie in their single-minded pursuit of hypertext, and Nelson discovered that Engelbart's hypertext project, NLS, was much more mature than his own work with Andy Van Dam.

A few months after getting together with Nelson, Engelbart prepared his work for its first public demonstration. On December 9, 1968, before a standing-room-only crowd at the Association of Computing Machinery Fall Joint Computer Conference, Engelbart, armed with his mouse and a graphical display terminal, took the computer science community on a ninety-minute tour of the future, showing them a fully developed hypertext system. He also demonstrated videoconferencing, conversing with his colleagues back at SRI. This demonstration marks a watershed in the history of computer science. The videotape is featured as part of the permanent collection of the Smithsonian Institution; over the following thirty years, nearly every aspect of Engelbart's research has been adopted as a basic feature of personal computing.

Though deafening cheers hailed Engelbart's work, it took years for its full importance to be recognized. The NLS remained a curiosity, an intriguing possibility that couldn't flourish until computers were both widely available and networked together. Very few people had direct access to computers in 1968, nor would they for another decade. Engelbart kept plowing ahead, harnessing a brand-new medium known as the Internet to expand the possibilities of intelligence augmentation, even though its utility remained restricted to a tiny community of researchers and government scientists.

Nothing could have been further from Ted Nelson's vision of a democratic medium for publishing, learning, and exploration of the wealth of knowledge, so, as the 1970s began, he became a hypertext revolutionary, gathering around him a coterie of fellow travelers, inspired by the vision of an ever-expanding universe of knowledge, instantly accessible and rich with links from subject to subject. Nelson found a name for this vision in a poem by Samuel Coleridge:

In Xanadu did Kubla Khan
A stately pleasure-dome decree:
Where Alph, the sacred river, ran
Through caverns measureless to man
Down to a sunless sea.

Henceforth, the project of Nelson's life—his Great Work—would be known as Xanadu. The pleasure-dome of the human intellect, caverns of knowledge measureless to man—all of these he promised his hacker elite, who spent years, first in New Jersey, then Michigan, and finally, California, working out the details for what would become both an elegant and universally inclusive system for the storage and retrieval of a planet-spanning repository of information. Nelson wrote and self-published a manifesto describing his vision and his efforts, titled either *Computer Lib* or

Dream Machines, depending upon from which side—front or back—you began reading. A weird, inverted text, it became an incredibly influential recruiting tool, infecting many of a new younger generation of computer programmers who were now building their own computers from cheap parts such as the brand-new microprocessors recently developed by Intel. In the few years before the personal computer revolution took off, a geek community of home-brew hobbyists took the computer out of the corporate data processing center, put it in their garages, and began to play. The possibilities seemed endless. So what to do first?

Nelson had a few ideas.

Copies of *Computer Lib/Dream Machines* were passed around the community like religious texts, closely read and deeply pondered. Nelson's troupe grew from a small band to a happy clutch of rabidly intelligent and eagerly enthusiastic minds, whose goal was nothing less than the transformation of the written word. They wrote philosophy. They wrote poetry. They wrote some computer programs. They came up with some amazing innovations, solutions to the problems inherent in creating a system comprehensive enough that it could actually store all of humanity's facts, fictions, sounds, and images. And they gave demos. Which is where I enter this story again.

Just a few months after I left MIT, the Xanadu crew and its associated fans—many of the same individuals intrigued by Eric Drexler's work on nanotechnology—converged on the MIT campus for Xanacon '82, a full weekend of talks about and demonstrations of the Xanadu system, which had just entered beta testing. ("Beta testing" means that a system is working, but is still having its problems ironed out.)

I arrived on campus just as Xanacon was breaking up, but I met a few of the folks who had worked on the Xanadu system. They thrilled me with their talk about the significance of their work: all information instantly available, everywhere, all the time, all woven into a seamless whole showing how each part depended upon every other. Proof that no one lived alone on an island of thought, but each related to everyone else in a great web of knowledge.

I couldn't make much of Xanadu's specifics on the basis of conversation, so I purchased a copy of Nelson's most recent text, *Literary Machines*. It blew my mind:

> Forty years from now (if the human species survives), there will be hundreds of thousands of file servers—machines storing and dishing out materials. And there will be hundreds of millions of simultaneous users, able to read from billions of stored documents, with trillions of links among them.
>
> All of this is manifest destiny. There is no point in arguing it; either you see it or you don't. Many readers will choke and fling down the book, only to have the thought gnaw gradually until they see its inevitability.
>
> The system proposed in this book may or may not work technically on such a scale. But some system of this type will, and can bring a new Golden Age to the human mind.

I didn't fling down the book. I got the concept. Ted Nelson was my new hero; all I had to do was wait for the release of Xanadu and the resulting dawn of the Golden Age. Of course, since I knew how to program, I began to think about how I'd want my own Xanadu to look. The Xanadu crew were only working on the servers, not on the browsers that people would need to get to this universe of information. As I waited for its imminent release, I began to design my own browser for the Xanadu system.

The Macintosh, released in 1984, provided what I believed to be the perfect platform for a hypertext system. Although I'd never heard of Doug Engelbart, many of his innovations made it to the mass market in the Macintosh. I could see the value of the mouse and windows in a hypertext system. So I read up on the voluminous Macintosh programming manuals, did a lot of planning, and, still waiting for the release of Xanadu, wrote a simple hypertext program for the Macintosh in the summer of 1986. I figured it would have to do until Xanadu became widely available, at which point I'd adapt my own work, making it a browser for the Xanadu system.

As I put the finishing touches on my own work to show to some venture capitalists (still a rare breed back in the 1980s), Apple released Hypercard, a hypertext system for the Macintosh. It wasn't Xanadu. It wasn't even as powerful as the system I was working on, but it covered enough of the bases to create a groundswell of interest in hypertext. From that time forward, I didn't need to explain what hypertext was; I just pointed to Hypercard.

Hypercard had one very serious drawback; a document—or stack, as it was known—could have links only to other pages—cards—in the stack. You couldn't link from one stack to another, which meant that the universe of links was very small indeed. This was perfect for a manual, for example, or a single book, but fell impossibly short of the dream of Xanadu, a dream I desperately wanted to see realized. It's only when links extend out from the computer into the world beyond that they become vitally important. And Xanadu had always been envisioned as a networked universe of text, images, and links.

Xanadu was never released. The pleasure-dome never opened. Nelson's project was acquired by software giant Autodesk in 1988, when it was nearly ready for release. But Autodesk's engineers insisted upon rewriting Xanadu from the ground up. It'll take us six months, they said, and set to work. Four years later, the engineers hadn't yet completed their rewrite of Xanadu, and Autodesk canceled the project. Nelson, who had signed away all control over Xanadu to Autodesk, scrambled to find another home for his Great Work. Fortuitously, on the day Autodesk canceled the project, Nelson had lunch with another hypertext true believer named Tim Berners-Lee, who'd spent the last decade working on his own system.

In 1980, Berners-Lee had never heard of Ted Nelson or Xanadu, nor Douglas Engelbart and his NLS. But he shared the same consuming passion, as he recounts in *Weaving the Web*:

> *Suppose all the information stored on computers everywhere were linked,* I thought. *Suppose I could program my computer to create a space in which anything could be linked to anything.* All the bits of information in every computer at CERN, and on the planet, would be available to me and to anyone else. There would be a single, global information space.

As a software engineer consulting for CERN, the European high-energy physics research facility located underneath the Alps on the outskirts of Geneva, Berners-Lee had access to some of the most advanced computers in the world, and because researchers needed to share their data, all of the computers at CERN were linked together on a network, enabling them to transmit messages to one another at incredible speed. He built a tiny program which he named Enquire, which allowed him to create little notes—like digital index cards—which could then be linked to one another, in as many ways as he thought appropriate.

For example, researchers were always passing through CERN, doing experiments, giving talks, working on fellowships, so Berners-Lee spent tremendous effort just to keep track of the scientists he worked with. Using Enquire, he could create a note with a researcher's permanent contacts, another with temporary contact information, yet another linking to relevant projects, and so forth, a tangled web of people and information that mirrored his own experience at CERN. And that's why it was useful: it reflected him.

His consulting contract soon came to an end and another four years passed before he managed to secure a fellowship at CERN. He'd been thinking about Enquire, had even tried to take it further with something called Tangle, a program that ended up becoming so tangled in its own logic that it crashed without a trace. Back at CERN he re-created Enquire and began to campaign for a more comprehensive tool to link all of the information in CERN's computers into a coherent, easy-to-access framework. Berners-Lee began to cast about for a name that could describe the comprehensive nature of his work:

> Friends at CERN gave me a hard time, saying it would never take off—especially since it yielded an acronym that was nine syllables long when spoken. Nonetheless I decided to forge ahead. I would call my system the "World Wide Web."

Even though these three words form one of the most widely known terms today, CERN officials remained unconvinced of the value of this Web. Most of the scientists at CERN—nuclear physicists—focused solely on their research experiments; few could see the value in any project linking their information together—particularly if it stole resources from their own investigations. But Berners-Lee found an ally in Robert Cailliau, who led the charge within the administrative levels of CERN to see that Berners-Lee got the funding and support he needed to complete his work. His design consisted of two equal parts: a web server, which allowed a computer to transmit a copy of any document it possessed, and a web browser, which allowed anyone at any computer to request documents from those servers.

Berners-Lee had long since decided that the Web would use the Internet—the then-tiny network of globally linked computers. Anything on the Internet could potentially be on the Web, and he thought that might go a long way to helping his tool take root. CERN, like Europe in general, had been slow to embrace the Internet, thinking it very American-centric. (It was.) But the Internet allowed very different computers to communicate in the same language—or protocol—which meant that the web server would have to speak only one tongue. Rather than learning the IBM language and the Siemens language and the NorskData language, all Web servers would talk to each other through the Internet. This may be the single wisest decision he ever made.

Berners-Lee and Cailliau took every opportunity to evangelize the World Wide Web at CERN, giving frequent talks, presentations, and demonstrations, doing their best to prove the worth of the Web to the ten thousand researchers and engineers, and slowly the pair made some headway. The Web solved a number of outstanding problems at CERN—how to create and manage documentation, for example—but it remained an underutilized tool, little more than a toy. The pair traveled to hypertext conferences—which had become potent academic affairs by the late 1980s—and found themselves snubbed. Everyone had their own approach, and everyone thought theirs to be the correct one. Berners-Lee didn't care much for the gimmicks and gizmos that made one hypertext viewer better than another; he just wanted to get all of the islands of information connected in the simplest manner possible. That made him the enemy of staunch academics, who had cooked up neat theories on the right ways to view hypertext or the right ways to store it. Berners-Lee was agnostic; he didn't care about the right way, as long as he could store the information and get it out again. That was enough for him.

As it turns out, that really was enough. His little programs were released to the public at no cost, as both program and source code, the lines of logical statements that comprised the program. They soon found their way into research labs in Europe and America, where they sparked a revolution in technology.

CERN, part of the global community of high-energy physics,

had a reputation that extended even into America's heartland. In 1992, at the University of Illinois's Urbana-Champaign campus, programmers at the National Center for Supercomputer Applications—or NCSA—took an interest in the Web. Aided by Berners-Lee's source code, a team at NCSA began work on a new version of his web server and web browser. The browser, named Mosaic, provided a very easy and elegant way to access the Web. Berners-Lee had done similar work, but had written his browser for a computer, called NeXT, which was years ahead of its time and never gained mainstream acceptance. The programmers at NCSA immediately brought the browser to the most popular workstations—high-powered computers used by scientists and engineers—making the Web visible to a much larger audience.

In the summer of 1993, I found myself at SIGGRAPH, the annual conference for computer graphics researchers and manufacturers. I wandered into SIGKIDS, an area specifically designed to showcase the possibilities of educational computing. All around me, kids played with high-powered computers, creating art, making music, or using an Internet-based videoconferencing tool to chat with folks in Australia. In one corner of the exhibit, I found a lonely computer. It had a single open window that said:

WELCOME TO NCSA MOSAIC

What is this? I wondered. I noticed that one of the words of text in the window had a blue underline, so I took the mouse in hand, moved it over the word, and click—I went to another page. Hmm, another hypertext system? I clicked around some more. Yep, it's a hypertext system. But there's not much here, is there?

It was well executed, this Mosaic thing, but I'd seen hypertext systems before. Heck, I'd *built* hypertext systems before. So I didn't think very much of it and wandered off to see something else. Silly me.

At that time I was using the Internet daily, both in my office at

Apple Computer and at home, on my brand-new-used Sun workstation. That was a ridiculously high-performance computer to have in my den, but it could connect to the Internet like nothing else, and I needed to be on the Internet. I'd been using the Internet since early 1989, and I really liked the ability to send a message to any of my other friends in the software biz, most of whom were connected at work. I could also read USENET, the gigantic bulletin board of Internet users, which covered topics as varied as vegetarianism or Zoroastrianism—something for everybody. And I could subscribe to mailing lists, receiving daily updates on subjects of interest to me. One of them, Fringeware, featured news of the bizarre, conspiracy theories too wild even to make their way onto *The X Files*. One day, one of the messages I received from Fringeware included this curious bit of text:

http://www.io.com/fringeware/index.html

What is this? I wondered. It was identified as a URL, but I had no idea what that was. But it did say something about Mosaic, which brought me back to that day at SIGGRAPH. The message helpfully included a paragraph telling me where I could get the Mosaic web browser. So, using the Internet, I downloaded the Mosaic web browser, then the Mosaic web server, and installed them on my overpowered home computer, which fortuitously happened to be the ideal computer for running Mosaic. I launched the Mosaic web browser, typed in this URL—and the universe changed.

I have no idea what I saw, but I remember going click, click, click, and traveling from one web page to another seamlessly, following the flow of links and ideas as I might wander around a library. I got it. *It was here.* After eleven years of waiting, it was finally here. Xanadu, or the Web, as they called it, was finally, absolutely, incontrovertibly here.

Picture me with a grin like a cat who has just eaten the world's tastiest canary. A grin that didn't leave me for two weeks, because, every night when I got home from work, I'd go online, launch Mosaic, and surf the Web. And at the end of that two weeks, I was done. *I'd surfed the entire Web.*

It might seem impossible to believe, but at one time the Web held only a few hundred sites and a few thousand pages; it's not that each of them was particularly significant, but the diversity of sites—weather, physics, MTV.COM!—told me that this was the real McCoy, something had come along which could hold *everything* in its grasp. The dream had become real.

I realized I had the Mosaic server software on my machine, and *Presto!*—my computer became server #330 on the World Wide Web, perhaps the first example of a home-based website. I went to the NCSA web pages and learned how to compose web documents in HTML, the strange language of brackets and quasi-English words that formats Web documents in such a way that they look the same from computer to computer. I toyed with creating a site, the Internet Arts Omnibus, or IAO, which I hoped would be a focal point for art on the Internet. (I hadn't seen nearly enough art as I'd surfed the Web.) Today we might call that an art portal, but back then it was just a wish, a possibility, a place for something great to happen.

I'd been seduced, converted, and I was ready to do some missionary work of my own. First stop was my mentor, Owen Rowley, who had worked with Ted Nelson at Autodesk as that company tried to rewrite Xanadu. I had him drop by, showed him Mosaic and all the cool sites I'd found through it, but he refused to believe that this weak cousin of Xanadu could amount to anything important. For a while, anyway. A few weeks later he got the Big Idea—I saw the lightbulb go on above his head—and he stopped in his tracks and said, "This is gonna be it, isn't it?" To which I simply replied, "Yes." "Okay, let's tell everyone."

At this time I lived in San Francisco, which, although located just above bustling Silicon Valley, had basically been untouched by the technological revolution happening beneath its feet. A lot of computer programmers lived in San Francisco and commuted to the Valley for their work, but very little of that culture rubbed off on the city itself; it was living in the gently fading glory of the post-hippie days. Only *Wired* magazine had put a stake in the ground, declaring San Francisco to be the epicenter of a revolution that had not yet happened. Everything else was as it had been for twenty years. That was about to change.

For that peculiar set of people who shared an interest in both technology and the arts, a regular series of parties became de rigueur. These were known as Anon Salons, basically rent parties for the less prosperous members of the community, events where people could come and show off their latest toys or newest tricks. Many of the folks who came to Anon Salon had been involved in some way with the virtual reality industry, which centered itself in nearby Marin County; others worked in Silicon Valley at places like Apple, Adobe, or Electronic Arts.

In early January 1994, I dragged my home computer across town (it weighed about as much as a boat anchor) to Anon Salon, strung a hundred feet of phone cable to the apartment's single jack (this being long before people regularly added a second line for their modem) launched my computer and gave a demonstrations of Mosaic. As the party began to fill with people, I'd spot someone I recognized. The pitch was always the same. "Hey, [insert name of friend here], you wanna see something cool?" "Sure, what?" "Name something you're interested in." "Hmm . . . Okay." They'd pick a topic, something like gardening. I'd click over to this little site called Yahoo!—still run from a lab at Stanford University—type in what they'd named as their interest, click on a link, and hear, "Whoa, that's cool! What is this?" Welcome to the World Wide Web, my friend.

It's useless to convince someone that a technology is important. No technology is important—it's the *use* of it that's important. I didn't try to tell anyone how revolutionary the Web was; instead, I showed them how the Web reflected them. I let the Web seduce them with its human face. For many of the folks in the high-technology community in San Francisco, this was their first exposure to the World Wide Web. I don't know if I had anything to do with San Francisco becoming the hotbed of the web revolution, but I'd like to think that evening's Anon Salon got a lot of people dreaming about the possibilities that had suddenly become real.

NOOSPHERE

We all know the rest of the story. Suddenly everything, *everywhere,* gets onto the Web. A few months later, a friend told me that he was setting up a web design company. What's that? I wondered. Why would anyone need web design? (I admit I'm dreadfully slow at times.) Well, he replied, we have some clients who need websites built for their companies. Really? Who? Volvo. Club Med. MCI. That sort of thing. Wow, I replied. There's a business in this?

Before business came to the Web (my friend was in the vanguard), a galaxy of individuals staked out their own little corners of cyberspace. Three hundred servers became three thousand nearly overnight, offering everything from pop star fan pages to pet care advice to alien abduction conspiracy stories. Yahoo! moved out of its offices at Stanford and rapidly grew into a hundred-billion-dollar behemoth. Some very creative scientists at Digital Equipment read the entirety of the Web and allowed you to search it through Alta Vista. At the end of 1994, the Web encompassed nearly a million pages, so a search engine wasn't just a luxury; the idea of surfing the entire Web became a ridiculous impossibility.

By the middle of 1995, three thousand servers became thirty thousand; now companies rushed to get onto the Web, creating "storefronts" and warehouses, promotions and pornography. But the popular groundswell, led by millions of people who saw the Web as their own playground, continued unabated. The Web grew weirder, wilder, and more a reflection of the diverse population who used it. A few clicks, perhaps a scan of a photo, some quickly typed text, and your unique perspective could take its place in the emerging crowd.

There's little question that America went web-crazy in the latter half of the 1990s; the Web became the most important story of the time, affecting every institution—social, commercial, governmental—that it came to touch. And it came to touch *all* of them. Anything unconnected became unimportant; anything on the Web transformed itself into a global resource, accessible anywhere, anytime, for any reason at all.

The flashy stories of overnight web millionaires (and billionaires) became the loudest notes in the symphony of the early web era, leading many to believe that electronic commerce was the reason for the Web; this was an easy mistake to make, because these newly rich could afford to advertise their status—in the hope of achieving even greater wealth. But these tales only partially hid a more fundamental change: Our relationship to knowledge became instantaneous, ubiquitous, and far more integrated into the fabric of culture than it ever had before. Want to know who composed the music for your favorite movie? Find out at the Internet Movie Database. How do you spell that word? Click to *Merriam-Webster's Collegiate Dictionary*. What's the gross national product of Argentina? Check the *CIA World Factbook* (on the Web since 1996).

It's as if a new layer of mind—collective intelligence—enveloped the Earth as the twentieth century drew to a close. In half a decade, the prosperous population of the Western world found itself drawn into cyberspace, finding prizes so alluring they became impossible to resist. (Of course, there was a fair amount of garbage mixed into that mound of jewels.) Even without promotion, without the endless hype or clever marketing campaigns, the Web would have been everything it is today because it needs nothing more than itself to be the all-consuming object of human fascination. It has become the ultimate seductive technology—because we all worked to build it, adding our own individual identities to the whole.

Other than a very few visionaries like Ted Nelson, who remains dissatisfied with the Web, no one foresaw the importance and comprehensive impact of the World Wide Web. But, over fifty years ago, one fairly obscure scientist did predict a coming transformation of the human mind, the birth of collective intelligence, and the emergence of a new way of knowing.

In the closing years of the nineteenth century, as a young boy on his father's farm in Provence, Pierre Teilhard de Chardin loved to study the earth. The rocks and mountains of the region formed the backdrop of his active imagination, together with frequent discoveries of fossils, ancient bones uncovered as teams tunneled through rock to lay railroads within Europe's mountain ranges. These fossils seemed to contradict the widely held belief that God had created the Earth only a few thousand years before, on October 22, 4004 B.C., the date calculated in the mid-seventeenth century by Irish bishop James Ussher.

Such contradictions didn't bother Teilhard one bit. A devout Catholic, he got a first-class education, courtesy of the Jesuits, and after graduation, he went on to become an ordained priest and member of their Society of Jesus, a militant wing of the Catholic church that advocates radical thought, held in check with a strong orthodoxy. The Jesuits have always been at the forefront of the sciences, while paradoxically maintaining their deep faith. This suited Teilhard perfectly; his Christian beliefs formed the foundation of a lifelong study of prehistoric life, as recorded in the fossil record, and led him to draft a comprehensive analysis of planetary evolution, culminating with a vision of the future of humanity.

After serving as a stretcher bearer in World War I (Teilhard received the French Legion of Honor for his work), he taught in Paris, but found himself drawn to the wastes of the Gobi Desert, an unwelcoming expanse of dunes and dust in China's northwestern corner. Located in the rain shadow of the Himalayas, the Gobi is perhaps the driest region on Earth, a perfect environment for the preservation of fossils.

Teilhard spent a decade in the Gobi, and traveled further afield, to Kashmir, Burma, and Java, examining the bones of the earth for clues to our beginnings. Through plain bad luck, he found himself trapped in Beijing as the Japanese attacked and invaded, forcing him to wait out the Second World War in China. Stranded half a world from home, this confinement left him free to organize and refine his data and theories about evolution and the history of the Earth. After the war ended, Teilhard returned to France, only to have his strange new ideas about planetary prehistory greeted with indifference. The Jesuits assessed Teilhard's work and deemed it too radical to be published. He accepted their dictates—one of the principle Jesuit ethics is obedience to authority—and moved to New York City, teaching and continuing his research until his death in 1955.

Unlike Galileo, who lived under a papal edict banning his writings, Teilhard was always free to publish his work, but he chose to obey his superiors' wishes. After his death, his research reached the public in several volumes. The most important, a summing-up of the natural history of the world entitled *The Phenomenon of Man*, warns in its introduction:

> If this book is to be properly understood, it must be read not as a work on metaphysics, still less as a sort of theological essay, but purely and simply as a scientific treatise.

Although Teilhard could have argued the merits of evolution from a purely theological basis—which makes him rare among Christian apologists—he chose to use the scientific evidence he had collected through his years of research to advance an astonishing proposition: man is absolutely the intended development of the evolutionary history of the Earth. Rather than being the glorious accident so many evolutionary biologists claim humanity to be, Teilhard stated that every moment of life on Earth led up to the creation of a species that could come to grasp the process—evolution—through which it had been created.

How can this be possible? Isn't humanity the accidental result of being in the right place at the right time with the right genes? This, Teilhard pointed out, flies directly in the face of the fossil record, which proceeds from the simplest single-celled organisms, such as algae, into more complex forms, such as bacteria. These sophisticated cells make the evolutionary leap into multicellular life forms, such as plants and jellyfish. Here, Teilhard notes, something unusual happens: once organisms become sufficiently large, they stop growing in size and turn some of that energy inward, becoming more complex. Now we see the birth of distinct organs, such as gills, the liver, and the heart; the living creature becomes a well-integrated collection of organs functioning as a whole.

These creatures, more complex than any that had come before, invaded the ecosystems of the Earth. The fish took over the seas; insects and reptiles dominated the land. Some of their interior development had given rise to their own awareness; a simple nervous system helped these complex creatures find food or respond to danger in a wide range of environments. Although unsophisticated, these first nervous systems gradually grew more complex, until, with the advent of the mammals more than 100 million years ago, the brain became one of the most massive organs, demanding a quarter of the body's energy just to keep running smoothly. These fast, tiny, warm-blooded mammals had the intelligence to evade the consequences of a massive catastrophe that wiped the large, slow, and small-brained dinosaurs from the surface of the planet, an environmental apocalypse that gave mammals the opportunity to flourish.

Apparently intelligence is its own reward; in a dangerous environment filled with predators, the smartest animal lives the longest, long enough to have offspring to carry that intelligence forward into another generation. Mammalian nervous systems grew in size and complexity, until with the primates—animals such as gorillas, chimpanzees, and humans—it took a quantum leap. Whole new sections of the brain evolved, and all of a sudden, these creatures inhabiting the trees and the savannas of Africa began to use simple tools, bringing their intelligence outside of themselves in forms that in turn helped them augment their own abilities—very much Engelbart's dream for the NLS, but ten million years earlier.

There is no clear dividing line between chimpanzees, gorillas, and humans. We know that as recently as six million years ago, very little separated us from either of these close cousins. Perhaps, as some scientists have recently proposed, the formation of Africa's Central Rift Valley separated identical tribes of primates; those on the wet western flanks evolved into chimpanzees, while those on the dry eastern plains, driven out of the trees to forage in a less hospitable environment, developed into the ancestors of humankind. As less capable members of the tribe couldn't survive to reproduce, these more gifted pre-humans were caught in an accelerating spiral toward intelligence.

This, too, Teilhard found absolutely predictable. For him, human beings represented the point at which the growth of the nervous system moves from external size to internal complexity, from awareness of the environment to consciousness of the self. This is the defining feature that separates human beings from other animals—our extraordinary degree of self-consciousness, which is the root of our languages, our religions, and our cultures. We see ourselves as different from the other animals because our self-consciousness has identified us as such.

With bigger brains and brighter minds, with history, which brought self-consciousness into the larger culture, human beings conquered the Earth. From our origins in Africa, wave after wave of our species spread out to cover the continents—first Asia and Europe, then Australia and America. Soon nearly every place that could support life supported *human* life; our intelligence allowed us to operate in every ecological niche, every climate, every environment—even upon the oceans.

As the pyramids received their finishing touches, the final wave of human exploration began. Tribes of seafaring Melanesians struck out in their canoes and catamarans toward the far-flung and fertile islands strung across the Pacific Ocean, the delicate chain of coral reefs and volcanic outcroppings known as Polynesia. Just a thousand years ago, the last Polynesian tribes settled on the islands they named "Hawai'i," and the age of migration came to its close. The Earth lay covered in a blanket of humanity.

We were fruitful, and multiplied. By the time Captain Cook "discovered" Hawai'i (where he would die, the victim of an unfortunate misunderstanding), the Hawaiian Islands were already home to nearly the same size population they support today. We colonized and then crowded Earth's surface, a singular mass of life.

Although the nations remained separate, the idea of humanity caught hold, a recognition of the reach of our species across the entire globe. Tenuous though it was, the fabric of humanity wrapped the planet in a new layer of self-consciousness, a recognition of the common origins of all humanity, a widely held metaphysical ideal that became a scientific theory when Charles Darwin published *On the Origin of the Species by Means of Natural Selection.* Teilhard saw the discovery of the processes of evolution as a culminating moment not just in human history, but in the history of life itself:

> The consciousness of each of us is evolution looking at itself and reflecting upon itself.

Teilhard marked no difference between our consciousness and the way we came by it; evolution has driven us to this point, and he insisted that evolution would carry us through. The accumulated mass of humanity, like so many separate cells, would soon turn inward, growing more complex. From the single consciousness of the isolated human, Teilhard believed a collective intelligence *must* emerge, a layer of mind covering the planet like a blanket of air: a noosphere (from the Greek *noos*, or mind).

Teilhard made few claims about the noosphere, but he refused to believe that any collective intelligence would make us less individual. He argued that the movement of life tends always toward greater differentiation, greater complexity, so he presumed that the noosphere would create a space of human possibility, rather than a homogeneous, stifling group mind. The nations might come together and the peoples of the world might unite, but the qualities that make them human, and thus unique, would only deepen and intensify.

The other quality Teilhard granted the noosphere was speed. Now that humanity had grown to cover the Earth, the next natural maneuver would be the emergence of an internal complexity, something new to connect the islands of human mind into a greater union. Once that process began, Teilhard thought, it would proceed quickly until every mind had been transformed. The noosphere would suddenly burst out of the shadows of thought to occupy center stage. It would be a transformation total and undeniable.

Which brings us back to the Web.

Never in the course of human history has any innovation

taken hold so rapidly or so completely. Entire nations—the Chinese, for example—find themselves threatened. Industries churn. Ultra-Orthodox rabbis ban its use for its "profanations." Yet it draws tens of millions of us in, day after day, offering an endless exploration of the catalogue of humanity, all its stories and wishes and horrors and facts. Every day it becomes more indispensable, more unforgettable, more important. And it's not even ten years old.

We can't know for sure if the Web is the same thing as the noosphere, or if the Web represents part of what Teilhard envisioned. But it feels that way. We do believe there is something new under the sun, as if something unexpected and wonderful (and ridiculous and horrible) has arrived. How else to explain this sudden transformation? If Teilhard was right, *the Web is part of our evolution*, as much an essential element of humanity as our acute eyes, our crafty hands, and our wonderful brains.

If Teilhard's predictions are true, there's a chance that his theories are as well. No one can say for sure; perhaps in twenty years it will be unmistakably clear. In any case, we will never again be in a world without the Web. It's revolutionary for us, because we straddle the gap between a world before the Web and a world never again without it, but it's entirely matter-of-fact to our millennial child. She has a new organ for thinking.

Even as the Web grows, it disperses invisibly into the environment, becoming part of the fabric of our world. Freed from the computer and its locked-down location, the Web can be found in cars, on cellular phones, even in refrigerators. It seems to have a drive of its own to inseminate everything with its presence, so that to be in the human world means to be linked to the Web.

Picture this millennial child hard at work learning about the world, and everywhere surrounded with the infinity of human experience. From her earliest days she has learned how to venture forth into that immensity, chart its currents, and build maps to the things she needs—be they facts, fantasies, or fears. For her, the act of knowing has become inseparable from the act of reaching for knowledge. She searches for what she needs to know; in a moment's time, the answers are at hand. And anything known to anyone anywhere has become indistinguishable from what she knows for herself.

8 MAKING PRETEND

PLAYING GOD

I've pulled myself away from my web browser, only to be drawn into another computer-generated diversion, this one on my television. I'm inclined to think of the television as an entertaining toy, while the computer monitor always reminds me of work. Even though the Web can be playful, it can't begin to rival the kind of fun I can have with my television; you see, I own one of the seventy-five million PlayStation video game consoles that Sony has manufactured since the toy's introduction in 1995.

In the late 1980s, after building up a reputation as *the* consumer electronics company, manufacturing televisions, VCRs, compact disc players, and stereo components, Sony decided that it would be best served by owning the entertainment that played through its products. They bought Columbia (formerly a part of CBS) and garnered an enormous entertainment beachhead in films, television, and music. Once Sony had integrated those pieces, the corporation that gave the world its first transistor radio turned its eyes upon a market even more lucrative than film: video games.

As the twenty-first century began, the video game industry, controlled almost entirely by three companies, accounted for over $7 billion in revenues in America, a fraction more than the film industry grossed during the same period of time. At $49.95 a pop, games are big-ticket entertainment, whereas Sony is lucky to collect eight bucks every time you walk into the multiplex. Of course a well-designed game will provide much more than two hours of entertainment—sometimes as much as fifty or sixty hours. And, occasionally, much more. As with the game I'm playing now.

I'm building an entire city. My official title is mayor of a place that I've named Markville. (What the heck.) Right now, Markville is little more than a patch of dirt on my television. But I have a budget, and I'm going shopping. Using my controller, the button-laden device that connects to Sony's PlayStation, I select some of the things I need. First, a bulldozer to clear the land. Maneuvering the controls, I work my bulldozer this way and that, turning the untrammeled grass into a strip of bare earth. Next, I raise a patch of clapboard houses, arrayed around a central street, though at this point, it's little more than a dirt path.

Soon enough, people will come to live in the newly developed Markville. They'll be drawn to the brand-new houses, to the open fields, to the opportunities that await them in the town, if I do my job right. All the while, a marquee at the top of my TV screen is giving me suggestions as to what to build next. Perhaps a general store? Or a few more houses? My decisions will have a direct impact on the livability of Markville. People won't live where they can't buy things, and without enough houses, the general store will go out of business. It's a complex system of dependencies, and every time I make a decision, I close out some possibilities even as I discover new ones.

Now a few hundred people live in Markville, and all of a sudden the disorders of a larger community begin to develop. I've passed a critical threshold in public safety and need a police station. Furthermore, I need to *pay* for the police station, which means I need to begin to tax the residents of Markville—a burden they had previously been spared. The residents grumble a little bit (their protestations make their way onto the marquee) but

Markville is safer as a result, and the overall satisfaction of the residents begins to climb.

If I'd let the situation spiral out of control, by building up the town without any of the services civilization needs to remain stable and fair, my population would likely have grown increasingly discontented. They'd clamor for law and order, and if I remained deaf to their repeated entreaties, I'd face a growing sea of troubles eventually culminating in a revolution, which would forcibly remove me from office. Game over.

The game I'm playing—SimCity 2000—does its best to create a realistic world, where actions have consequences. It's the latest of several generations of SimCity products, the brainchild of programming wizard Will Wright, who in the mid-1980s decided that simulations of the real world could be just as compelling as any fantasy shoot-'em-up or casino simulation. Fifteen years of solid sales have proven him right; today, over seven million copies of SimCity run on nearly every computer imaginable, including the various video-game systems. In 1997, Wright's company, Maxis, was purchased by gaming giant Electronic Arts for $125 million, ensuring that new generations of SimCity will continue to be released.

In 1993, as the Clinton administration began to push on a national health care initiative, Wright wrote and released SimHealth, designed to demonstrate the complexities, difficulties, and pitfalls of health care in America. Though not wildly successful, SimHealth did demonstrate that a computer could be a powerful tool for examining the interconnected array of issues confronting a society.

Since their invention, computers have always been used for simulation. The very first computers, invented in the 1930s, were used during World War II to decrypt the Nazis' secret codes, simulating the operation of the Nazis' famed Enigma encryption machine. By the end of that war, the Allies knew all of the deepest secrets of the German high command and could easily outfox their enemy, even sending them counterfeit coded messages that looked absolutely authentic. Just as the war ended, the first digital computer, ENIAC, entered operation. A thousand times faster than earlier analogue computers, ENIAC was used to simulate the trajectories of artillery shells, a task that had previously occupied legions of human computers, and performed a day's worth of calculations in about fifteen seconds.

By the early 1950s, IBM had begun to manufacture computers (even though company president T. J. Watson had predicted that the total market for these machines was no more than five or ten units) and found an eager customer in the U.S. military, who wanted to use computers to simulate the strategic threats of the cold war. Using a network of spy planes and listening posts, U.S. military intelligence gathered a constant stream of data on the Soviet military. This information was fed into IBM's computers, which created detailed models of the possible threats to American security.

A decade later, an MIT graduate student named Ivan Sutherland would make computer simulations personal. Working late at night on the TX-2, an early computer equipped with a display and a light pen (a forerunner of the mouse), Sutherland created Sketchpad, the first interactive computer program. Using the light pen, Sutherland could draw on the display, creating the first computer graphics. It was all very straightforward—and no one had ever seen anything like it before.

In 1964, Sutherland left MIT for Harvard and began work on a computer system that could display three-dimensional graphics, images with depth. Although the small, black-and-white computer displays didn't reveal the full glory of these three-dimensional images, Sutherland had another flash of inspiration: why not create a display using two screens, precisely directing information into each eye? Our perception of depth comes from the fact that our eyes receive slightly different images; from the discrepency in these images, our brains sense the relative distances of objects in our view.

The head-mounted display (HMD), Sutherland's solution to the 3-D problem, looked either very Space Age or utterly ridiculous, depending upon whom you ask. It fit over the head like a bicycle helmet. On either side of the head, tiny displays projected images into the eyes. When these displays were connected to Sutherland's new computer system—voilà!—images would appear before his eyes in three-dimensional simulation. Sutherland published his work in 1968, the same year that Douglas Engelbart introduced NLS, but like the NLS, it was too far ahead of its time. It gave computer scientists something to dream about, but remained an intriguing impossibility for all but the very few labs dedicated to computer graphics.

Four years later, one of Sutherland's students, Nolan Bushnell, would turn the quiet world of computer graphics inside out. As a college undergraduate, Bushnell had seen a program known as Spacewar, developed at MIT in the early 1960s. It used a computer monitor to simulate the battle of two spacecraft flying above a planet. Spacewar had been programmed to simulate real space flight, with the same laws of gravity that a spacecraft might encounter in outer space, and was so engaging that it had become an underground sensation in computer science labs around the world.

Bushnell, a big fan of Spacewar, spent years trying to design a cheaper version (it ran on million-dollar computers), but try as he might, he couldn't do it. He gave up on Spacewar and tried his hand at something simpler—a computerized game of Ping-Pong. He quit his Silicon Valley job and set to work. A few months later, a prototype of his computer game, named Pong, showed up in Andy Capp's, a watering hole favored by many of the engineers employed by Silicon Valley firms. Although just barely cobbled together, Pong was a huge success: by the end of the evening everyone in the bar had put at least one quarter into the game. And at ten A.M. the next morning, people were lined up outside Andy Capp's to play Pong! Atari, the company that

Bushnell founded to build his video games, became one of the fastest-growing companies in American history. Computer graphics had entered the mainstream of popular culture.

Over the next decade, Atari would redefine games and introduce the public to the idea of interactivity. On the way, it would swell to a multibillion-dollar enterprise, finally crashing as competitors gobbled up the market it had created. By 1985, Atari would be a mere tenth of its former size. At the same time, just a few miles away from Atari's headquarters, another generation of research scientists were developing the next revolution in computer simulation.

Tucked away in a corner of NASA's Ames Research Center, the same facility where, at the dawn of the twenty-first century, nanotechnology research is now taking center stage, scientists were taking Ivan Sutherland's pioneering 3-D work to a new level. In the 1960s computers were barely past the transition from vacuum tubes to transistors. By 1984, researchers could build systems using integrated circuits holding up to a million transistors apiece. Sutherland's multi-ton computer was reduced to a few pounds. His displays, which had used bulky cathode-ray tubes, had been replaced by liquid crystal displays (LCDs), just a few inches in size and weighing ounces. Everything that Sutherland had demonstrated could now be designed in the form of an efficient, *wearable* unit.

The Virtual Environment Workstation (VIEW) project at Ames connected a high-performance computer to a new, lightweight, head-mounted display. The computer could create entire scenes with a fair degree of realism, and the head-mounted display provided a way for a VIEW user to look around within the synthetic world, unconstrained by any attachments to the computer. You could even walk around inside the simulation as the computer tracked your movements.

One thing distinguished the VIEW system from anything that had come before it, taking it beyond even Sutherland's dreams.

An enterprising Stanford medical student by the name of Tom Zimmerman invented a device that allowed the computer to accurately track and respond to the movements of the hand. Known as the Data Glove, it was a spandex mitten with bend sensors wired over the joints of the fingers; when you moved a finger, the computer could sense the precise movement. When the computer was supplied with the position of your hand and fingers, it became possible to put them in the synthetic world.

Thus was virtual reality born.

Sutherland had supplied the eyes to peer into the synthetic world, but Zimmerman provided the interface—a way to interact with the synthetic world that was as natural as interacting with the real world. The Data Glove closed the gap between seeing and doing, between mind and hands. You could reach out and touch things that did not exist. Yet they seemed virtually real. The realm of imagination had suddenly become tangible.

The researchers at Ames used the VIEW system to design a virtual wind tunnel. Within the virtual environment, you would see a model of the space shuttle—just then preparing for its first flights—surrounded by colored lines, representing the flow of air around the shuttle's surfaces. With the Data Glove you could reach into the environment and reshape the air flow or even reshape the shuttle, feeling your way to the appropriate configuration of wings and airflow to produce a stable and reliable flight.

Before VIEW, such a task was laborious and boring; a researcher might sit in front of a computer terminal, typing in various values, watching the simulation results change. All of a sudden, this became more like a game than like work. Getting your hands into the simulation meant that you could observe changes and respond to them immediately. VIEW made an abstract engineering task visible and visceral; you could almost feel the shuttle in your hands, responding to your computerized caresses.

The seductiveness of virtual reality is that it makes the relationship between humans and computers completely natural. Rather than acting as an ethereal thinking machine, the computer spends its energies expressing itself sensually, in terms that we're well equipped to understand, thanks to four billion years of real-world evolution. Our bodies have been blessed with a computing ability that greatly exceeds even the most sophisticated of today's machines: we can look at objects and identify them, listen to language and understand it, feel a surface and identify its qualities. All of these seemingly simple tasks are impossible for a computer, which means they really aren't so simple. Our brains have learned to make sense of our complex and constantly changing world, a world very immediate and real. When researchers at Ames, following Sutherland's lead, built computers that could present themselves to us at that level, through our senses, we were suddenly able to perform even the most complex tasks nearly effortlessly.

A number of people immediately recognized the importance of virtual reality. One was a former Atari programmer named Jaron Lanier, who formed VPL, designing virtual reality systems for clients like Matsushita and Boeing. Lanier had a knack for publicity, and as he toured the country, lecturing at universities and museums about virtual reality, public interest in this newest technology of computer simulation began to reach a fever pitch. It seemed a given that virtual reality was going to be the next big thing.

MAKING CYBERSPACE

At this same time I was making a comfortable living as a software engineer at a Cambridge company founded by a pair of MIT graduates. Shiva Communications made devices that connected to Macintoshes and their networks. What made these devices unusual was that you could use them to get access to your office network from anywhere in the world. Using your Macintosh and a modem—a device that had become a lot more common over the decade—you could dial into the network, and your computer could use the network just as if you were in the office, to receive and send files, read electronic mail, and so forth.

Today we think nothing of being able to perform such feats; back then, Shiva was practically the only company offering such products. During my job interviews with the founders, one of them expressed a vision: "Wherever there's a phone jack," he said, "that's a connection to the network." It impressed me—the implications were staggering—so I took the job.

I was charged with writing the software for network management. In layman's terms, this means that I would write the programs controlling the Shiva products connected to customer networks. We packaged this software with our products and prided ourselves on the fact that it was so easy to use, thanks to the Macintosh's intuitive interface, that our clients rarely had to open a manual.

The Macintosh was unique in that it was designed from the start to communicate over a network. Networks allow computers to exchange any sort of data between them, but in the days before the World Wide Web, this was generally documents or electronic mail. Some of our clients, like behemoth Hughes Electronics, had thousands of Macintoshes on their networks, and even our intuitive management tools couldn't handle all of the sophisticated configurations they'd set up in their offices spread out along the coastline of southern California.

We decided that I should create a new generation of tools that would be just as intuitive to our smaller clients, but offer incredible power to our largest users. As I sat down to design this tool—Shiva NetMan—I confronted a problem that had plagued software engineers for decades: what does a network look like?

That's not an easy question to answer; networks are for the most part invisible, silently passing their data back and forth across their spans. To map the network, that is, to make it visible, I'd have to apply some arbitrary system of rules, which would define what the features of the invisible network looked like. I thought about this for a long while and plotted out several solutions, each of which was unsatisfactory. The possibilities that might satisfy our smaller clients wouldn't handle the complexity of our larger customers, while anything sophisticated enough to satisfy the big guys would simply confuse the little folks. I knew that I wanted a "one size fits all" solution, but I couldn't visualize how that might be done.

Shiva closed down for the last week of 1990, and in the span between Christmas and New Year's Eve, I polished off some reading that had been sitting around for a few months. One of the books, which I had been meaning to read for some time, was William Gibson's science fiction novel *Neuromancer*, highly recommended by a close friend. Deep in its pages, I came across the solution to my problems:

> "'The matrix has its roots in primitive arcade games," said the voice-over, "in early graphics programs and military experimentation with cranial jacks." On the Sony, a two dimensional space war faded behind a forest of mathematically generated ferns, demonstrating the spatial possibilities of logarithmic spirals; cold blue military footage burned through, lab animals wired into test systems, helmets feeding into fire control circuits of tanks and war planes. "Cyberspace. A consensual hallucination experienced daily by billions of legitimate operators, in every nation, by children being taught mathematical concepts. . . . A graphic representation of data abstracted from the banks of every computer in the human system. Unthinkable complexity. Lines of light ranged in the nonspace of the mind, clusters and constellations of data. Like city lights, receding."

Gibson hadn't puzzled through the difficulties of making the invisible visible—he just *did* it, painting a picture in words that anyone who understood the problem would immediately appreciate as the perfect, elegant solution. *Make the network a place,* Gibson suggested, *and they'll know what to do.* Cyberspace was the answer.

Such an undertaking was admittedly ambitious. Shiva, a start-up company with only sixty employees, couldn't afford to underwrite my adventures in science fiction made fact. I continued to work on NetMan, but my heart wasn't in it. I was hungry for cyberspace. I found everything I could on the subject of virtual reality and digested it. Sometime in January, one of my employees presented me with an issue of a magazine that had just been introduced. *Mondo 2000* touted itself as the hip guide to the future, with a banner on its cover reading: "The Rush is on! Colonizing Cyberspace." This looks auspicious, I thought, and dived in, soon coming across an interview with VPL founder Jaron Lanier:

> Virtual Reality is not going to be the television of the future. It's going to be the *telephone* of the future. And that's the key thing.

As I read these words, the world went white for a brief and utterly magnificent moment. Although I had understood Gibson, only when I read Lanier's words did I grasp the full import of virtual reality. Beyond my own ideas for an intuitive system for managing invisible networks, Lanier prophesied a world where human communication took place in the infinitely imaginary worlds of virtual reality, where the medium of communication would become as important as whatever we communicated through it. VR would inevitably become a new language, unbounded by anything except our own ability to use it.

It felt like a religious experience and fundamentally changed the direction of my life. There was no going back. Try as I might, I completely lost interest in my work. But the fortunes of the company depended upon Shiva NetMan, so I dove back into my work, finished the project, and began to plot how to make a career in virtual reality. There was no other option.

I began to design a virtual reality system from easily acquired parts. A motorcycle helmet and twin LCD pocket televisions, together with a few lenses, some cardboard, a hefty amount of duct tape, and I had built a head-mounted display of my very own. True, it was both heavy and more than a little claustrophobic, but it worked. I got two video cameras, mounted them onto a tripod, cabled them into my home-brew head-mounted display, and suddenly I could put my eyes anywhere the video cameras could go. I had a friend wheel them around my apartment as I went on a virtual tour. I dreamed up a business opportunity where I would lead the charge on the next generation of video games, using virtual reality.

I was hardly alone in thinking that video games were the next great frontier for virtual reality. From its earliest days, VR entrepreneurs had been working very hard to bring this idea to the marketplace. Working with Mattel, Jaron Lanier had adapted his $10,000 Data Glove, turning it into the Power Glove, a $100 accessory for the enormously popular Nintendo Entertainment System (NES). Although the idea of the Power Glove sounded impossibly cool to a generation of youngsters, translating a $10,000 device into a $100 toy meant that the resulting toy, while workable, wasn't either particularly exciting or fun. To use the Power Glove, you had to keep your hand extended, away from your body, for long periods of time. This is simply impossible for anyone to do. The Power Glove was a cool idea that no one had actually tried to use before they marketed it. The "hottest toy in the world" that year came flooding back to Mattel, returns of a product that would never sell—except to a few hard-core hobbyists like myself, doing some down-home VR research.

I focused on the head-mounted display, the essential element in creating a visible virtual reality, and continued working on software that could network tens of thousands of players together. By now, I had enough technical background in both of these areas to believe I could make a good run in business, so in the fall of 1991, I moved to San Francisco, the ground zero of the VR revolution. With some partners, I started a company named Ono-Sendai, after a fictitious firm in Gibson's *Neuromancer*. Ono-Sendai was the manufacturer of cyberspace decks, devices that plugged into the global network and generated the visual representation of cyberspace. That's what we were going to do.

From our garage workshop, we began to design hardware and software that—I believed—would inevitably take the world by storm. Along the way we tried to sell our ideas to venture capitalists. They'd chuckle and say, "You're going up against Nintendo and Sega? They'll kill you." It wasn't quite clear if they meant that as literally as it sounded. Although they agreed our ideas were solid, they unanimously considered that any frontal assault on the two giants of video games—Sega having swept into prominence with their Genesis video-game system, toppling Nintendo's supremacy—would be nothing less than economic suicide.

By the fall of 1992, running low on money, we decided to dance with the devil. I scheduled appointments with executives at both Sega and Nintendo to show them our work. Sega was located just a few miles away in Redwood City, while Nintendo's corporate offices were across the street from Microsoft's, outside of Seattle, so we met with Sega first. On December 11, 1992, I showed them the results of two years' worth of work. They seemed almost disinterested in the presentation but, at the end of the meeting, they asked us to sign an agreement that would allow them to tell us what they'd been working on, so long as we agreed not to tell anyone else. We signed it and learned that Sega had been working on their own VR system for the past several months. Sega Virtua VR, as they called it, would be a complete head-mounted display for the Sega Genesis video-game system. They were planning to preview it to a few select toy buyers at the

Consumer Electronics Show, the enormous annual trade show taking place in Las Vegas about four weeks after our meeting.

Sega had a problem, though. Their system didn't work. In particular, the head-tracker—the device that lets the computer know which way your head is pointing so that it can send the appropriate images to the display—wasn't performing properly. This is an absolutely essential part of a VR system; if the head-tracker doesn't work correctly, the system is likely to cause virtual motion sickness. Real-world motion sickness is caused by a confusion of signals delivered to your brain through your inner ear, the home of your sense of balance. If a VR system tracks your head movements too slowly—that is, if it delays more than a tenth of a second in reacting to your movements—it can make you sick. That might be virtual motion sickness, but the vomiting it induces is very real. I had designed a head-tracker as part of Ono-Sendai's efforts to produce a low-cost head-mounted display. It turns out that this device (which is now patented) was one of the most important pieces of work I've ever done. It was very inexpensive and it worked, unlike the designs Sega had produced. They were stuck, just a month away from the big show, with no product.

We walked out of that meeting with a deal to integrate my designs with theirs. If it worked, they'd license our design. This was big business: Sega planned on selling at least a million of the Virtua VR systems the following Christmas. It would be next year's "hottest toy in the world." And Ono-Sendai would be getting a piece of every one.

All through that holiday season, through Christmas and New Year's and into the next week, I worked eighteen-hour days, writing programs, testing them, fixing them, trying to ensure that my device would perform as advertised. Finally, just a few hours before the show opened, we attached our sensor to their prototype (using a fair amount of the ever-present duct tape) and Sega opened the room to a select few invitees.

It worked. Although cobbled together out of bailing wire and chewing gum, it actually worked. Sega seemed satisfied and told me they wanted to take my designs into production. Lawyers conferred, and soon we had our licensing agreement. We were going to be rich—or so I thought.

As Sega produced the very first finished prototypes of the Virtua VR system, they sent a few of them off for testing to SRI—the same think tank where Doug Engelbart had developed the NLS thirty years before. Sega wanted to be sure that the device wouldn't hurt any of its users, who were likely to be boys between the ages of ten and eighteen. SRI ran a few months of tests and returned the data to Sega, who kept the results secret, then quietly killed the entire Virtua VR project. (This was difficult, because they'd already rolled out the full court press, garnering the June 1993 cover of *Popular Science*.)

When SRI was informed they couldn't release the results of their studies, they agreed, but then did an identical set of studies on their own dime. They released these findings to the public.

It turns out that using a head-mounted display for long periods of time (over twenty minutes) causes your eyes and brain to adapt to the computer-generated images. These images aren't perfect replicas of those in the real world. Consider the following: if you hold a pencil close to your eyes, you'll need to focus on it to see it well, and everything farther away will be seen as a blurry object. In head-mounted displays, all objects, whatever their "virtual" distance, remain in constant focus all the time. That's because the objects are being projected from screens only a few inches away from your eyes. You can project slightly different images into each eye to counteract this effect as VR systems do, but that's only one of the six very unique ways your brain uses to tell you how far away something is. And that was the trouble; these VR systems were cheating the brain, but weren't doing it nearly completely enough. That lie was causing the brain and eye to become confused. The confusion would remain, even after the head-mounted display was taken off.

How long your brain remained confused was an open question. It varied from person to person. Some people adapted back to the real world almost immediately. Others seemed to take hours. No one knew what kind of effect it would have on growing children, who have highly flexible nervous systems. Could long exposure to a head-mounted display 3-D environment cause permanent damage? Even today, no one knows for sure (certainly, no one wants to do such a study) but the open question of permanent brain damage was enough to scare Sega away for good.

That was the death knell for Ono-Sendai. Sega had become our lifeline, and once the ties were severed, I had to close up the business and figure out what to do with my dreams and my ideas. The head-mounted display I had helped to design turned out to be dangerous—an awful but important lesson, and the reason why head-mounted displays are rarely seen, even today. But I still had a few tricks up my sleeve: I'd always seen VR as the perfect networking technology, and now I could drop everything else and focus on that.

Two years earlier, before I left Cambridge for San Francisco, I'd had an insight. Call it the six degrees of software. I'd realized that only by sharing information could really large cyberspaces, like the "Matrix" described in William Gibson's novels, ever hope to exist. No single computer could ever hold all of cyberspace. (Imagine trying to contain the entire Web on one computer!) So all of the computers, working together to create the consensual hallucination of cyberspace, would have to adopt a common understanding—a protocol—that would allow them to ask questions of each other about what was where in the visible landscape of cyberspace. (This is analogous to the "six degrees of separation" principle, which states that you can reach any person on Earth by connecting to one person, who connects you to another, and so on, until you reach the person you're looking for.) I called this idea "cyberspace protocol," and had already spent many months studying both the mathematics of three-dimensional graphics and computer communications, looking for some way to marry them. When solved, I believed I'd have something that everyone using VR would want as part of their projects, an indispensable connection between all the islands in cyberspace.

Until the late 1990s, each VR system contained its own tidy little universe, but each of these universes was entirely separate from all others. You might be in cyberspace when you voyaged through one of these worlds, but it was a very lonely place. You were the only one inside it. Even very sophisticated systems could contend with only two or three people. The U.S. military was building systems that could accommodate a few hundred people simultaneously, but I was thinking even bigger—much bigger. Someday soon, I imagined, there will be millions of people on the Internet, and each of them will want to be able to help shape the collective imagination of cyberspace. We needed a system designed with the capability to handle more information than anyone had ever imagined being available. After many months' work I'd developed the mathematics for Cyberspace Protocol, and on my newly acquired Sun workstation I slowly programmed my way into it.

At the end of September, I had a very primitive version of Cyberspace Protocol up and running. Now I needed to show some folks, get their comments, and figure out what to do next. By late 1993, nearly every VR company had gone out of business. The market in virtual reality had proved itself to be as virtual as its namesake. There were products, but no buyers. I was sure that VR would rise again (and it has, in video games) but I had no idea how to promote my own work without an industry to support it.

My programs were intelligible only to a few mathematician friends. They could verify my results; that at least reassured me that my math was solid. But looking at lists of numbers (the results of my programming) wouldn't impress anyone lacking a rigorous background in the mathematics of 3-D. To show people what my work meant, I needed to build an entire VR system. And while I'd been working in VR for a while, I certainly didn't know how to produce a complete system. Sure, I could learn, but that would take months, maybe years. And I was growing impatient. The Web, which I'd explored by now, had gotten me all fired up. It seemed as though Big Things were going on, and I wanted to play my own part in the unfolding drama.

As 1993 drew to a close, I met two individuals who would become close friends—and who would supply the missing pieces to my software puzzle. Servan Keondjian is a mathematician, a British-educated physicist who'd spent most of his youth playing with computers. In the early 1980s, the BBC had run a very popular television series called *The Computer Programme* about, of all things, computer programming, all based around an inexpensive Acorn computer, which was for sale across Britain for about $350. Millions were sold to a generation eager to explore the new frontier of computing. Britain is a programming powerhouse today mostly because of the BBC's early educational efforts. Unlike American computers of the time, such as the Tandy TRS-80 or the Apple II, the BBC Micro had substantial graphics capabilities. The generation who grew up programming these machines developed a nearly instinctive understanding of the complexities of computer graphics.

Years later Keondjian realized that he really wanted to translate his understanding of physics—how things behave in the real world—into computer simulations. In doing so, he could create virtual worlds that followed the same rules as the real world. This was a holy grail for Keondjian and many others working in computer simulation. The first step toward realizing his vision was the creation of a renderer, a bit of software which draws 3-D objects on the computer screen based upon their mathematical representation within the memory of the computer. A renderer is a serious piece of software because it must be both realistic—that is, it must create realistic-looking images—and it must be *very* fast. A television screen, for example, redraws its image thirty times a second. Each frame follows another in close succession, creating the illusion of movement. A renderer must do much the same thing, redrawing objects on the display at least twenty times a second if people are to believe that the objects on the screen represent reality, and not just some disorienting high-speed slide show.

In about nine months Keondjian had created software he called Reality Lab, a rendering program designed to be both fast and capable of representing the physical world. Reality Lab did what most engineers believed impossible: it took a run-of-the-mill IBM PC and turned it into a virtual reality system. Suddenly, a $1,500 computer had the same 3-D graphics power as a $100,000 computer workstation, and all sorts of new possibilities opened up.

I didn't have a clue how to write a renderer; when I met Keondjian and got a demonstration of Reality Lab, I begged him for a copy. I told him what I was trying to do. Although I couldn't afford the $5,000 license fee he'd asked of his other potential clients, he seemed very interested and told me he'd think about my request. Two weeks later, a package with some diskettes and an alarmingly brief manual appeared in my mailbox. Keondjian had sent me a version of Reality Lab and his blessing to use it in my quest for cyberspace. That solved half of my problem; I was about to find my answer to the other half.

Tony Parisi is seven days older than me and, as it happens, we're fairly alike in both temperament and interests. Through mutual friends in Cambridge, we arranged to meet up when he and his wife, Marina, moved to San Francisco in the closing days of 1993. On New Year's Day 1994, as I watched them unpack their lives into their new apartment, we had a long, rambling discussion about my ongoing work in virtual reality. I mentioned that I had all the pieces I needed save one, something called a parser. A parser is a computer program that translates bits of text into commands that can be understood by the computer. A parser would enable the computers I was connecting together to communicate with one another about what was out there in cyberspace. How could they tell each other about the shapes and forms in their individual little universes? A single computer understands everything within its own memory. But once data leaves the comfortable insides of that machine and heads across a network to another, possibly very foreign machine, the two computers need to have some sort of agreement about what shape that data represented. How do I tell another computer that one object is a green box, and that another one is a red sphere? I understood the theory behind parsers, but I'd never built one. (I'd dropped out of MIT before I could take a course in parser design—one of the great gaps in my own education, at least by my estimation.)

When Tony heard this, he laughed. "Mark," he said, "parsers are what I do for a living." I adopted my begging posture immediately. (I'd gotten used to it.) Tony didn't need much convincing. He'd been fascinated by virtual reality and seemed intrigued. In retrospect, Tony says he did it mostly to humor me. It was just a few hours of work, but a few hours of work for him would have meant months of study for me. A few days later we sat in Jumpin' Java—one of San Francisco's ubiquitous coffeehouses, strategically located between our residences—and set to work. In an hour's time I'd outlined what I wanted, scribbled across several pages. A few days later we met again, and he gave me a disk. "Here it is," he said. "Make it work."

I now had my hands on all three pieces: Servan Keondjian's renderer, Tony Parisi's parser, and my own Cyberspace Protocol. It took about a week to knit them together, and then I added one more feature. I wired my Cyberspace Protocol into the World Wide Web. Just as Tim Berners-Lee chose to connect the World Wide Web to the Internet, I connected my own work to the World Wide Web. This minor addition made it possible to see images on the World Wide Web in 3-D.

In the beginning of February, I gave Tony Parisi a call. "Come on down. It's done." He practically flew over, and ten minutes later, I gave him the first demonstration of the result of our work. At my computer, I had Tony launch the NCSA Mosaic web browser and go to a site I had added to the Web. It opened to a page with a single sentence on it. "To see the demonstration, click **here**." He moved the mouse over the word *here*, and clicked on it. For a moment, nothing happened—the computer was thinking—then another window opened in front of the web browser. There, in the solid black of empty cyberspace, was a single item—a banana!—floating gracefully in the void. Unlike a normal web browser, which pictured a flat, two-dimensional page, this window opened into a three-dimensional virtual world. Using the mouse, Tony zoomed up to the banana until it filled the entire window. He smiled and said, "Pretty cool."

That was the first part of the demonstration. "Put your mouse over the banana," I suggested. He did so, and the mouse pointer changed into a pointing finger—just as it would when it passed over a link in the browser. "So I should click on this?" he asked rhetorically. He did, and—very quickly—another window opened up. This time it was NCSA Mosaic and a web page I'd written. "Congratulations! You're back on the Web!" Tony had just gone from the Web into three-dimensional cyberspace—*our* cyberspace—and then, from there, back into the two-dimensional Web. The world of William Gibson and the world of the Web were joined forevermore.

I emailed Servan Keondjian, thanked him profusely, and reported our success. A day or two later, contemplating what to do next, I idly surfed the web pages put up by Tim Berners-Lee, the father of the Web, and saw that on one of them he wrote about the importance of virtual reality. Berners-Lee had been caught up by the allure of VR and felt it would be an important addition to the World Wide Web. He indicated that anyone conducting research in this area was welcome to contact him. Immediately, I wrote a note, thanking him for his incredible work on the World Wide Web and reporting our own successes on combining VR with the Web.

A response came in a few hours later. Would I be interested, he asked, in presenting my work at a conference he was just now planning? The First International Conference on the World Wide Web, to be held at the end of May in Geneva, the home of CERN, would gather web researchers from all around the world for a week of papers, presentations, and more informal "birds of a feather" sessions. CERN could accommodate about 350 conferees and Berners-Lee secretly feared that he might not get even half that number. The Web was still very new and almost entirely unknown.

And so I soon found myself on Air France, enjoying sumptuous food and flying toward Geneva. Although Tony had wanted to attend, he had a full-time job and couldn't drop it to fly halfway across the world for a conference on a still obscure networking technology. I was on my own.

Between the time that Tim Berners-Lee announced his conference and the conference's start date, the Web mushroomed in popularity. It was being used at nearly every university in America and Europe and people were discovering new ways to put it to work nearly every day. Suddenly Berners-Lee faced an entirely different problem: over 800 people wanted to attend a conference that could comfortably handle less than half that number. He closed registration and waited for the conference to begin. (At least a hundred unregistered conferees sneaked into the opening session.)

Those five days at CERN were some of the most fantastic of my life. Berners-Lee gave a stirring opening speech, calling for a "Bill of Rights" in cyberspace. Clearly he thought there was something to protect. A system for digital commerce was demonstrated: electronic money on the Web. Then the presentations began—over fifty of them. It was impossible to attend everything, to see everything, to learn everything, so I focused on the presentations of greatest importance to me—in particular, anything to do with media (images, sounds, even a museum!) on the Web.

Along with the full series of presentations, the "birds of a feather" meetings—gatherings of individuals interested in a particular topic—took place nearly round the clock. On Tuesday morning, I crowded into a small CERN conference room with Tim Berners-Lee, Dave Raggett, and a number of other researchers to discuss practical VR applications for the Web, things that could be done in the short term. What would we need to bring VR to the Web?

Raggett insisted that we'd need a language to define three-dimensional virtual worlds on the Web. As one of the fathers of HTML, he understood the importance of standards. HTML guaranteed that Web documents would look precisely the same from computer to computer, no matter where you viewed them. Something very much like that would be needed for VR to work on the Web. He called it VRML, a virtual reality modeling language (after HTML, which stands for hypertext markup language). I told him that such a thing already existed—I'd be showing it during my presentation, on Friday morning.

I'd named my VR web browser Labyrinth after the myth of Theseus and the Minotaur, but Dave Raggett had given a name to what Tony and I had invented. We had created the first version of VRML. I quickly reworked all of my presentation, excising "Labyrinth Scripting Language" from my slides and replacing it with the initials VRML. I hoped to unfurl the VRML banner during my talk, something I could call others to rally beneath. I was realistic: Tony had a full-time job, I was living off of an occasional consulting contract, and there was no business here, no way to keep a brand-spanking-new VRML alive. We needed friends—preferably, some powerful ones.

Fortunately, the attendees at the First International World

Wide Web conference met this criterion. These would be the people who would define the direction of the Web over the next few years, individuals like Brian Behlendorf, a nineteen-year-old dropout from the University of California who had just been hired to maintain the computers for *Wired* magazine's forthcoming website, *Hotwired*. Behlendorf had studied engineering before he'd left Berkeley behind and had just completed a course in computer graphics. He had no trouble seeing that VRML could become an important part of the Web. (Behlendorf was the friend who later told me he was setting up a web design business—a remark I greeted incredulously.)

On Friday, I gave my presentation before a large audience. I remember one academic who came up to me afterward, muttered, "Very impressive," then fled the room. Later that day, over beers in a downtown Geneva brasserie, Behlendorf offered his help. He claimed he couldn't offer much, just some space on the *Wired* web server and the ability to set up an electronic mailing list. I'll take it, I replied. As it turns out, that was all I needed.

I'd been very clear with Tony Parisi about one thing: we shouldn't try to own VRML. I was still badly burned by the failure of Ono-Sendai, and of VR in general. I believed that the rabble of tiny VR companies, each trying to rule the world by refusing to hang together, were inevitably hanged separately. The Web, on the other hand, seemed to march to a different beat. Everything on the Web was about sharing—codes, tips, sites—and it seemed to me that much of the Web's success sprang from this collective effort. That was what VRML needed.

Tony had no financial illusions where VRML was concerned, so he agreed that we could share our work freely. We both knew that if VRML proved popular, we'd still be able to make some money from it—creating VRML products or providing VRML services. (Which is exactly what happened: Tony later started Intervista, a VRML tools company, while I wrote books on

VRML and taught college courses.) I took the goodwill generated at the conference and used it to establish a small electronic mailing list—just the folks I'd met in Geneva—and through them I sent the word out to the world: If you'd like to participate in a project to bring VR to the Web, join us!

Our work had been a proof of concept, workable but not really ready for prime time. Neither of us knew enough about graphics. We had relied on Servan Keondjian's Reality Lab to handle the dirty work. I knew a lot about networking, and Tony knew a lot about computer languages, but that was only a beginning. We needed to get the expertise of other folks who had done similar work. And we were under a deadline: I'd promised Tim Berners-Lee that we'd have a specification—a complete, detailed definition of VRML—by the next Web conference, scheduled for Chicago in just five months.

I sent out the call on a Sunday evening in early June, and expected at most fifty researchers would chime in, adding their energy to our efforts. Monday morning, when I examined my electronic mail, I received a bit of a shock. Hundreds of requests to join the mailing list poured in from all around the world. By the end of the week, nearly two thousand individuals had accepted the challenge. VRML was on its way.

In retrospect, our actions were totally sensible, especially if you remember your childhood fables. In the story of "Stone Soup," a town experiencing a famine gets an unwelcome visit from a band of soldiers. Hungry, they're told there's no food to be had. One of them fetches a large cauldron, fills it with water, and adds a few stones. "Mmm, mmm," he says, "this is going to be the best batch of stone soup ever!" The townspeople regard him as half-mad, but he manages to talk one woman into giving him a few potatoes she has hoarded away. Then another woman offers up some onions. Another gives him some cabbage. And so on. Finally, the town butcher contributes a duck. Soon the "Stone

Soup" is a luxurious stew, which the soldiers and townspeople enjoy together.

I'd set my own cauldron of water to simmer in 1991; contributions from Servan Keondjian and Tony Parisi got the stew going, and, by the summer of 1994, we found a butcher to throw a chicken into the pot. That was Rikk Carey, a manager at Silicon Graphics—*the* home of 3-D computing—who offered up Inventor, a computer language that described three-dimensional graphics. Inventor had been around for a few years and was used by engineers at respectable corporations such as Ford Motors. Inventor had a pedigree, as it came from the acknowledged masters of computer graphics. Carey proposed using Inventor, with a few modifications for the Web, as the basis for a VRML specification. We came to a rapid agreement, and Rikk assigned two of his engineers to work with Tony and me on the drafting of a VRML specification. We completed our work four days before the Chicago conference. There, our presentation of the proposal was accepted, and VRML—version 1.0—was on its way to becoming the standard for 3-D on the World Wide Web. Tony and I had taken the seed of an idea, and with lots of help, we had gone the distance, from vision to working prototype to international standard.

As we caught our breath, we asked ourselves a more important question, one that Tony and I had both been avoiding all this time: what can you do with 3-D on the Web? We had some vague ideas. I was in love with the idea of VRML, but hadn't thought much about what I'd *do* with VRML once it was complete. Now, with the specification complete, this was the obvious next question.

Several months later, I'd see something that would rock me to my foundations, and convince me that I'd found the perfect application for VRML. Its very reason for being.

SEEING GOD

By June of 1995, quite a few people were working in VRML. Rikk Carey headed a crew at Silicon Graphics developing the first VRML web browser. Tony Parisi had gone off to found Intervista and began his work designing VRML tools. I was nearly finished with a textbook on VRML, written in the hope that I could teach people to create their own VRML worlds. Our efforts would be seen by the public for the first time when the Interactive Media Festival took over Los Angeles' Variety Arts Center.

As I toured the festival, a lot of folks were already inside, chattering about this exhibit or that. I ran into a friend—a Hollywood movie director—who asked me, "Have you seen T_Vision?" He led me up the stairs—and into another world.

On a five-foot television screen directly in front of me, I saw an image of the Earth in space, floating gently against a sea of black. Beside the screen, a large ball—looking like an overgrown beach ball—sat in a metal frame. Someone went up to the ball and gave it a spin. The earth moved. It was 1:1. As you spun the ball, the image moved on the screen.

Okay, that's nice.

Someone else walked up to the ball and pressed some buttons mounted on its frame. Now the image of the Earth began to fill the screen. As it grew closer, I began to see the individual features of the Earth—continents, then lakes, then rivers. We got closer and closer and *closer*, and the closer we got, the more I saw. Now I could make out cities and towns and industrial belts. We were zooming in on Germany, then on central Germany, then on Berlin; then I could begin to make out the streets, and closer in, I could see the buildings, and finally, just one building; and then, with a screeching halt, I was poised outside the window of an office building, looking in.

This is *very* nice.

We rode the trip in reverse: from building to streets to city to landscape to continent and, just as suddenly, a serene Earth floated pleasantly on the screen. I'd ended my journey where I'd begun, in high Earth orbit.

My friend leaned in. "That's not all. It's *live*."

"What?" I croaked. This was too much.

Someone must have heard us, because another button was pressed, and—voilà!—the cloud patterns, *live* via satellite, appeared over the image of the Earth. I felt like I was in the space shuttle looking down. No, it was more than that. My heart was full. I felt like I was seeing God. It was perfect. I had dreamed that such a thing might be possible someday, but never in my wildest moments did I imagine that it would be here, *now*, right in front of me, *live* and in color. I had to know everything about it, so I approached a tall gentleman working that amazing device and asked him to tell me the story of T_Vision.

In 1988, Joachim Sauter, a professor at Berlin's exclusive Academy of Fine Arts, founded the Berlin design firm of ART+COM with almost no money but with ideas galore. Located in the heart of old West Berlin, ART+COM brought artists and engineers together around a series of projects intended to demonstrate the coming integration of computers, communication, and design. At the time, few imagined that the Internet would sweep the planet, and no one knew that the World Wide Web would soon provide universal access to the global information resources of humanity, so the ultra-staid German business establishment had practically no use for such a renegade group of dreamers. If businesses couldn't see the value of the coming convergence between design and communications, Sauter decided he'd create a few pieces—artworks—which would at least shock and entertain the public.

One of Sauter's first installations, titled *Zerseher* (a pun that roughly translates as "disceiver"), portrayed a famous Renaissance painting, *Boy with a Child-Drawing in His Hand*, by Francesco Carotto. The original is striking—some might say strange—as the child pictured wears an almost demonic grin. When you walked up to Sauter's copy of the painting, that odd grin might cause you to stare at it for a few seconds. If you did, you'd see something very curious happening: wherever you looked at the painting, that part of the image would smear, then disappear. The more you looked at the painting, the more you destroyed it. Eventually, the devestated image would be replaced with a pristine copy and the process would begin again.

Sauter used some high-tech magic to accomplish this feat. Behind the projected image, which looked very much like an actual painting, a video camera tracked the faces of people who walked up to the pseudo-portrait. Using software written by an ART+COM colleague, Sauter examined the video to locate human faces. From that bit of information, he could then fix upon the eyes and learn where people's gazes were pointed. Then he'd erase the painting at that location, continuing the destruction as their gaze moved about the image. The longer you looked at the painting, the more it disappeared.

It was all very subversive, and as *Boy with a Child-Drawing in His Hand* is the first painting known to depict a child's artwork, Sauter thought it a very clever commentary on the infantile state of computer graphics at the beginning of the 1990s. As with most works of modern art, some people loved it, but many others hated it.

Zerseher got ART+COM some badly needed recognition, and some even more badly needed contracts. Berlin, formerly two separate cities, wanted some help planning the proposed reconstruction of the eastern downtown—Potsdammerplatz, the heart of old Berlin. The entire area, mostly filled with ugly Communist-era offices, was to be leveled, and a brand-new city built in its place. But what would that look like? The city planners turned to ART+COM, and asked them to create a virtual tour of Berlin—yesterday, today, and tomorrow.

Using old photographs, movies, and maps, the ART+COM

crew, led by Sauter, meticulously reconstructed prewar Berlin in virtual reality. You didn't need a head-mounted display to explore this vanished world; ART+COM projects shied away from the gear associated with virtual reality, favoring wall-sized displays. With the exception of the depth perception associated with head-mounted displays (the source of their dangerous effects on the nervous system), the effect is fairly similar. You could stand in front of a wall and stroll through a Berlin that had been bombed out of existence almost fifty years before.

Next, they re-created the present-day Potsdammerplatz, which would soon disappear forever. Then ART+COM worked with the city planners to place models of the proposed buildings into the simulation. Now the bureaucrats could actually see how Berlin would look in ten years' time, when all the construction would be finished. With the Berlin Wall torn down, they dreamed of a city that harkened back to its prewar glories, but still pointed toward its future as the capital of a unified nation. Using the simulation created by ART+COM, the city planners could gaze upon a future Berlin before the first day of construction had begun.

All the while, the staff at ART+COM were learning how to use their computers for ever more impressive tricks. They had a good selection of high-end Silicon Graphics computers for their work—the de rigueur tools for virtual reality—and were now learning how to create enormous virtual worlds, with thousands of virtual buildings covering hundreds of simulated square kilometers. It got them to thinking: if we can create a model of Berlin, can we follow it up with something bigger?

At just this moment, Deutsche Telecom came calling on ART+COM, asking them to develop a project for the ITU Conference in Kyoto, Japan, six months hence. Sauter offered them an idea so over the top that, if it worked, Deutsche Telecom would have the most impressive demonstration ever given at the ITU. (The folks at ART+COM weren't completely sure they could make it work, but what Deutsche Telecom didn't know wouldn't hurt them.) Those six months went by quickly in a blur of designing and programming. When the fateful opening day arrived, nothing in Kyoto was going as planned.

Pavel Meyer, dark-haired, short and compact, was sweating, swearing, and tapping on a keyboard. Next to him, Axel Schmidt, a tall blond who rarely loses his composure, looked worriedly into a computer monitor. Here in Japan, both men were working as if their lives depended upon it. Sauter paced back and forth, making occasional circles around Gerd Grüneis, who looked unconcerned. But he had started chain-smoking, lighting a new cigarette as he tapped the previous one out. Two hours to go and the brilliant idea wasn't working.

Meyer and Schmidt were crouched in front of the twin monitors of a million-dollar Silicon Graphics Reality Engine, a top-of-the-line supercomputer designed to make short work of even the most complicated computer graphics. That is, if it's programmed correctly. The bigger a computer is, the more likely it is to be finicky about its programming. Reality Engines were notoriously finicky. They could create breathtaking images, but they had to be coaxed—nursed—into doing what was asked of them. And this particular Reality Engine was being asked to do something only one other computer had ever been able to do.

Meyer was at a loss. The program had been working fine when he'd boarded a plane to Kyoto, the ancient capital of Japan. True, he'd cut it close to the wire (it had only started working an hour before he had to leave for the airport), but Meyer had felt sure that everything would go smoothly when he installed the program upon his arrival.

All around them, the final preparations for the International Telecommunications Plenipotentiary Conference were almost complete. The ITU, a branch of the United Nations, holds a conference every four years, a sort of Olympic event gathering all of the biggest telecommunications companies together under one roof for a week of policymaking, parties, and demonstrations of Things to Come. With the reunification of Germany, Deutsche Telecom had become the largest telecommunications company in Europe, and at the ITU conference they intended to flex their muscle with an eye-popping demonstration of the future.

With less than an hour to go, even Grüneis started to show a bit of apparent nervousness. Although Meyer had left Berlin with a fully functional version of their ITU Deutsche Telecom demonstration, he had been testing his software on a different machine, seven thousand miles away. Perhaps this Reality Engine had a subtle difference in its configuration, something that would take Meyer and Schmidt hours to find, if they could find it at all. Deutsche Telecom had set the machine up for ART+COM as directed, but sophisticated computers often take weeks to tune. Each one has its own personality, its own settings and peculiarities. That was the likely problem here. Or so Grüneis hoped.

Sauter took to tapping his fingers on the gigantic beach ball interface to T_Vision. Neither he nor Grüneis could do more than look on as the two programmers searched for a solution to their problem. In order to make this project work at all, Schmidt had written a graphics code that squeezed every ounce of performance from the Reality Engine, and Meyer had built an ultra-fast database containing billions of characters of data—a very tall order for the early 1990s. In every way, they'd pushed the machine to its limits.

Now Sauter had to contend with Deutsche Telecom executives, who had suddenly begun to look very concerned. The doors are opening in half an hour, one said. Sauter could do little more than shrug and apologize. We're working on it. Just a little more time. The executive didn't look any less worried after Sauter's reassurances.

With less than ten minutes to go, Meyer found the configuration error, corrected it and with that T_Vision was up and running. Perfectly. For the next five days, conferees at the ITU

Plenipotentiary Conference crowded the Deutsche Telecom booth, marveling at this demonstration of German ingenuity. ART+COM had given a gold medal performance at the Telecom Olympics and had also spawned a technology that will be a critical part of twenty-first century life.

A month later, in Los Angeles, T_Vision made its American debut at the Interactive Media Festival. That's where I introduced myself to Joachim Sauter—modest, but clearly glorying in the reception the work was receiving. He accepted my praise thankfully, and, that evening, I met him and Gerd Grüneis at the Festival's bar to talk about T_Vision. They told me that their model of the Earth was composed of around five billion characters of data, that it could display everything on Earth's surface down to a few kilometers, and—in a few sections—down to street level. I learned that T_Vision was designed to be networked, that each T_Vision system (there were three in the world—Berlin, Kyoto, and Los Angeles) shares information with the others to create a more complete view of the whole from its parts.

That was an idea I could get behind. It sounded very much like my own work. It made me want to put a question to them: Have you heard of VRML? Yes, they'd seen it. Very nice. Do you think T_Vision could work in VRML? They laughed. We're pushing a Reality Engine to its limits with T_Vision. Unlikely that VRML could handle the job. Hmm. Well, maybe someday? Someday, sure.

That was enough for me.

In early 1996, I spent a week huddled over my computer. I unplugged the phone and didn't even bother to check my ever-increasing flow of electronic mail. I searched across the Web for real-time images of the Earth from space, and I found them at a curious site run by John Walker, a software multimillionaire who had earned his fortune as cofounder of Autodesk, an early developer of computer-aided design (CAD) software. Walker had recently retired to Switzerland at the ripe old age of forty-five, and, from his comfortable chateau, created an extensive website to showcase his personal interests. Among these, he included both astronomy and Earth visualization.

Nestled deep within Walker's site, I found that he obtained the photographs generated by weather satellites, prettied them up, and then posted them on his website. They were accurate to within an hour. I set to work, and, a few days later, I'd finished my own homage to T_Vision, which I named WebEarth.

WebEarth is a three-dimensional, live model of Earth, just like T_Vision. Unlike T_Vision, it doesn't have the extensive detail provided by a high performance computer. Anyone with a computer and a modem can access WebEarth. I wanted it to become universally available, so I made sure that it wouldn't overwhelm the computer of anyone who gazed upon it.

When I completed my work, clicked on a link in my web browser, and actually saw the Earth—*live*—floating in the inky black of cyberspace, I broke down. (WebEarth has had this effect on several people I know. Something about the image is startling.) It was beautiful—perhaps not quite as beautiful as T_Vision, but beautiful nonetheless. And anyone could get to it, anywhere. You can too: just go to http://www.webearth.org and see for yourself. It's the Earth, live and in color, and reason enough, I believe, to see the Web in 3-D.

Back in Los Angeles, T_Vision garnered the $5,000 judge's prize at the Interactive Media Festival, the first of many awards it would win in coming years as computers grew smaller, faster, and ever more capable with 3-D graphics. Incredibly, a million-dollar Reality Engine of 1995 would shrink to become the $300 video game system of the year 2000. Once that happened, T_Vision could enter any home, not as a demonstration intended for the technological elite, but in the familiar form of a wildly popular toy.

THE (NEXT) HOTTEST TOY IN THE WORLD

The evening of Friday, March 3, 2000, was a chilly one in Tokyo. Temperatures hovered near the freezing point. That's not so unusual—Tokyo sits at the same latitude as New York City—but what made this night different were the lines of people waiting outside the stores of Tokyo's world-famous Akihabara district, the "Electric City" that is home to most of the city's numerous electronics retailers. Akihabara is truly futuristic; here you might have purchased the first CD players, DVD players, and high-definition televisions as they went on sale. During business hours, it's always crowded: on a Saturday afternoon you can barely make your way through its narrow streets.

On this Friday night, all of the stores were closed, shuttered until Saturday morning. Yet people still gathered. From all across Tokyo's sprawling center and suburbs, subway trains disgorged legions of young people at Akihabara station—mostly young men, few older than age twenty-five—who joined the lines. They spent the night chatting excitedly. The revolution was coming—and they'd be the first to take part in it.

This revolution had been unveiled nearly a year before at the Computer Game Developers' Conference (CGDC), a yearly retreat for the software engineers and designers who create the game titles for Nintendo, Sega, Sony, and the PC. The CGDC is usually a low-key affair, almost a family reunion—most of the folks involved in the development end of the game industry know one another. But in March 1999, no one had been prepared for what they were shown.

For months, rumors had swirled through the game industry about Sony's successor to its incredibly successful PlayStation video game console. By 1999, Sony had sold over sixty million of them, decisively conquering the video game industry in Japan, America and Europe, and sweeping both Sega and Nintendo, once thought indomitable, into the periphery. True, Nintendo had fought back aggressively with their Nintendo 64 game console, in many ways superior to the PlayStation, but it seemed that Nintendo's secretive corporate culture had finally gotten the best of them. Game developers found Sony a pleasant and easy-to-work-with partner, very different from Nintendo. PlayStation titles filled the market, while the N64 struggled to keep up. As the average video-game buyer cares more about the variety of titles available for a console than about its design, the public flocked to the PlayStation.

Sega had not stood still, planning a return to prominence with their Sega DreamCast, which would up the technological ante in video gaming. The DreamCast used the extremely high-powered computer at its core to draw images on the screen at an unprecedented rate. In addition, the DreamCast was equipped with a high-speed modem so that video gamers could dial into a Sega gaming network and play with other gamers all across the world. Slated for release in September of 1999, DreamCast was expected to cut a swath through Sony's growing video-game empire. But on March 18, 1999, those hopes were dashed.

During the opening keynote of that year's CGDC, Phil Harrison, vice president of research and development at Sony Computer Entertainment—the division of Sony responsible for the PlayStation—told the game development community that Sony was planning to release a new generation of the PlayStation. Called the PlayStation 2, it would be so powerful that it couldn't be compared to anything that had come before, from anyone.

Game developers had heard this kind of hype before ad nauseam. It always accompanied the release of a new platform, and it hadn't ever been true. But Harrison continued. We have some video-game footage generated by the PlayStation 2 that we'd like to show you, he said. The lights went down, the footage ran, only a few minutes' worth. When the lights went up again, Harrison was greeted with a stunned silence. Then began endless, thunderous applause. The next stage of the revolution had begun.

Virtual reality had already started to find a home in some video-game consoles. When Sony introduced the PlayStation in 1995, it used 3-D graphics *exclusively*. All of the games written for it, such as Mortal Combat, a fighting game, and Grand Turismo, a racing game, created virtual worlds with characters and cars that looked fairly realistic. These games simulated the real world of physics—turn a corner too fast in Grand Turismo, and you'd flip your car. This was a big leap over the Super Nintendo video-game console, which used two-dimensional graphics for its games. The PlayStation felt more real and the public enthusiastically adopted it.

The essential element of 3-D graphics is known as a polygon. Visualize a cube—it has six square sides, or surfaces. But for the computer, that cube is composed of six polygons. All objects in 3-D simulations are drawn using polygons. The more polygons you can draw, the more realistic the world you can portray.

The PlayStation can draw about 360,000 polygons a second. That's pretty good, but remember that the PlayStation needs to redraw these polygons continuously to simulate action—thirty times a second to work on a television screen—so this means that the PlayStation's virtual worlds can only be composed of about 12,000 polygons, each redrawn 30 times a second.

The earliest virtual worlds had only a few thousand polygons in them and required the resources of hundred-thousand-dollar computers. Ten years later, Sony was designing chips for the PlayStation that outperformed these systems and was selling everything for $199.

Alvy Ray Smith, another pioneer of computer graphics, once noted that if you wanted to create reality in virtual reality, images that would look indistinguishable from the real thing, you'd need about 80 million polygons a second. Obviously the PlayStation fell far short of that. PlayStation games still looked cartoonish, but offered enough of an improvement to sell millions of video-game consoles.

After the triumph of the PlayStation, Sony polled their developers. What would you need for truly incredible games, if you could have anything you wanted? Well, they answered, could you give us 18,000 times more polygons? Sony checked with its engineers. Nope, we can't do that. Would you settle for 300 times more? If we have to, they smirked. But is that even possible? Wait and see, Sony replied. Wait and see.

With Harrison's announcement, the waiting was over. The PlayStation 2 promised to revolutionize video gaming completely, bringing unheard-of power into the home. The engineers at Sony promised that the PlayStation 2 could deliver an astonishing 66 million polygons a second, very nearly the number needed to create "reality" in virtual reality. Ultimately, the PlayStation 2 would likely peak at about 20 million polygons a second—a figure ten times greater than Sega's DreamCast, if not quite enough to re-create reality.

Sony didn't need to put out any hyperbolic press releases to promote the PlayStation 2. In May 1999, at the Electronic Entertainment Exposition in Los Angeles, crowds waited in a long line to see the preview models of the PlayStation 2 in its first public display. I waited along with everyone else for a chance to see this amazing device. What I encountered made my jaw drop. Computer simulations of faces that looked so human they might have been captured on film. Racing cars that looked like they were running the genuine Indy 500. It was incredible. The million-dollar Reality Engine couldn't create simulations this compelling, and yet the PlayStation 2 would sell for just $300.

Following the standard pattern of releasing the new video-game consoles in their home market first, Sony prepared to launch the PlayStation 2 in Japan. The release date, March 4, 2000, gave them ample time to fix any bugs that might appear before the planned American launch date of October 26, 2000. (As it turns out, this was a wise decision—early PlayStation 2 components were recalled.) Japanese consumers would pay a premium to be first—a PlayStation 2 would cost them nearly $100 more than Americans would have to pay. Despite the inflated price tag, Japanese gamers prided themselves on the fact they'd be the first to enter into an entirely new era of video gaming, the near realism of PlayStation 2.

Sony announced that they planned to sell a million units in the first weekend of the PlayStation 2's release, an unprecedented figure that underscored their confidence in their own ability to produce such a high-technology wonder. The corporation revealed it had designed two new integrated circuits, the Emotion Engine and the Graphics Synthesizer, for the PlayStation 2. The Emotion Engine is a very high-powered microprocessor, while the Graphics Synthesizer handles the work of drawing those 66 million polygons to the screen every second. As described, both chips are far more complicated than even the most sophisticated processors created by Intel, whose Pentium III had become the workhorse of the mainstream PC market. How could Sony pull off such an engineering feat? Many video-game industry pundits openly suggested that Sony simply wasn't up to the job. They'll miss their ship dates—we won't see the PlayStation 2 until 2001. It's just too complicated. Too many things can go wrong.

Sony never commented on these rumors beyond assuring their partners that everything was going as planned. As the millennium came and went, Sony prepared for the product launch. In February, a two-day PlayStation 2 Festival held just north of Tokyo gave the Japanese public its first look at the new system and at all of the games that had been developed for it. Hundreds of consoles crowded the exhibition floor, proof positive that Sony could manufacture the units in volume. The new games looked stunning, whipping the public's excitement to a fever pitch.

On February 14, Sony unveiled a new PlayStation 2 website to take pre-orders. Why stand in long lines when you can get one directly from Sony? Good idea, but Sony found themselves victims of their own popularity. In the first minute, the website processed *100,000* orders, then crashed. Sony engineers brought the site back up an hour later, and at the rate of 50,000 an hour, orders poured in. Three weeks before its introduction, the PlayStation 2 had already become the fastest-selling consumer electronics product in history.

For those unwilling to wait while Sony mailed them their PlayStation 2, there was always Akihabara. Over that long, cold evening, the crowd swelled. By Saturday morning, at least 10,000 people had queued up in front of Akihabara stores, waiting for 9:30 A.M., when they'd open their doors to the onslaught of anxious buyers.

By 10:30 A.M., not a single store in all of Japan had a PlayStation 2 left. Sony had shipped at least 700,000 units to retailers all across the country, and every one of them had been purchased. This was in addition to the 800,000 pre-orders taken through their PlayStation 2 website. The frenzy was so severe that *The New York Times* reported the sad tale of a sixteen-year-old who got about a hundred yards from Akihabara before he was robbed of his PlayStation 2 by two motorcycle-riding youths!

Sony had created the hottest toy in the world.

The design of the PlayStation 2 speaks volumes about the direction of the future of electronic entertainment. As the most sophisticated video-game platform ever created, the PlayStation 2 is designed to become the cornerstone of the family entertainment center. While it can certainly play games, the PlayStation 2 can also play DVDs (a format Sony invented). It has a curious selection of connectors, the kind normally seen on a garden-variety PC, but never before built into a video-game console. The PlayStation 2 has two universal serial bus (USB) connectors; these are used on PCs to attach things like a keyboard, a mouse and an image scanner, as well as more sophisticated items like a

CD burner, which can create CD-ROMs. It also has an iLink port, Sony's name for the ultra-high-speed network jack showing up on digital camcorders. Sony has said that they hope to allow PlayStation 2 owners to connect their digital camcorders to the unit so the owners can bring images of themselves into the system, then apply them to the game characters. This means that, in the near future, your video-game characters can *look just like you*.

The iLink port presents another even greater opportunity; it's possible to attach a disk drive to iLink, giving the PlayStation 2 the ability to access a huge volume of information. Although the DVDs the PlayStation 2 uses contain as much as five billion characters of information, that information can't be changed. When you attach a disk drive to a PlayStation 2, you now have a place to store dynamic data, information that changes over time.

Why might you want that? Consider this: your family buys a PlayStation 2, a keyboard, a mouse, a disk drive, and a printer. Now you have a fully functional home computer, for a fraction of the usual cost. Plus, it connects to the TV set to play DVDs, which also saved you a couple of hundred bucks for a separate DVD player. What you've got—what Sony has sold you—is a complete home computer. The only thing missing is some way to get it on to the Internet.

But Sony hasn't forgotten about the Internet. The final connector on the PlayStation 2 is a PC connector, used in laptop computers for things like modems and network adapters. Sony announced that in 2001, after the worldwide launch of the PlayStation 2 is complete, they will provide high-speed access to the Internet, using an inexpensive component that will fit into the PC connector. Suddenly, the PlayStation 2 was revealed to be a full-featured computer, only faster, cheaper, and with a better connection to the Internet than most home computers.

Threatened by Sony's growing strength, in March 2000 Microsoft unveiled its own answer to the PlayStation 2, the X-Box.

Slated for release in September 2001, its press material promised performance as much as ten times greater than that of the PlayStation 2. If so, this device would cross the final hurdle for virtual reality, the culmination of forty years of making pretend, a machine capable of generating images indistinguishable from reality. The incredible power of the PlayStation 2 opens the door to entirely new types of computer simulations, works of art, education, and forms of play that we've only just begun to imagine. In the early years of the twenty-first century, as we explore virtual reality in our living rooms, we'll begin to discover the power of making pretend.

9 THE REAL WORLD

SPIRITS IN A MATERIAL WORLD

Although the Web may not present the same range of entertaining possibilities as the Playstation2, it can still tickle my funny bone. There are still some genuinely weird sites on the Web, even in an era when everybody seems to be trying so hard to be practical and lucrative. A friend recently passed along a URL, so I've launched my browser and carefully typed it in: http://ouija.berkeley.edu. I'm greeted with an animated flickering candle and a title that identifies this site as Ouija 2000.

A new oracle for a new millennium? Perhaps. Unlike most oracles, this one gives detailed, polite instructions:

> **Before playing, please follow these 4 simple steps:**
>
> **First, gently move your keyboard to the side of your screen.**
> (And so I do. Click.)
>
> **Now place your mouse and mousepad in front of the screen.**
> (My mouse and pad now sit directly before my screen. Click.)
>
> **Now turn off the lights in your room or office . . .**
> (It takes a moment, but I dim the lights, pull the blinds, and sit in the monitor's glow. Click.)
>
> **Last, place both of your hands lightly on your mouse . . .**
> (Done. Click. Now I read a page of final instructions.)

> When you press the button below, a new browser window will pop up. The pop-up window has 2 smaller windows: one shows a live video image and the other contains the planchette, the oval shape that will move with your mouse.
>
> The planchette in the video window will also respond, but more slowly, as its motion depends on the actions of other players and on the influence of spirits in the system. Concentrate on the question at the bottom, rest your hands lightly on the mouse, and let your movements express your responses.
>
> Please note: many people feel that Ouija boards summon powerful forces. You have been warned.

That last line gives me a moment's pause; after all, one shouldn't invoke what one can't control—that's the lesson of the sorcerer's apprentice. But I gather my courage, make a final click!, and a new window opens up. The upper half contains a video image showing a planchette, the familiar heart-shaped plastic oracle, resting atop a Ouija board. The lower half shows an animation of my mouse, modified to look very much like the planchette. It moves about as I lightly push the mouse this way and that.

Suddenly, the live video image shows movement and the "real" planchette moves about, cruising the board's letters, displayed in a semicircle, finally coming to rest as I stop my mouse. Somehow, the vibrations I deliver to my mouse have been translated to actions on this distant Ouija board. It's not traditional magic, of course; I can spy a span of metal bolted onto the distant planchette, which is obviously moving it about. I can't see the other end of that metal bar, but it must be connected to a computer attached to the Web, a computer that reads my own mouse movements and translates them into the actions of the planchette.

At the dawn of the twentieth century, as the world became more mechanized and rational, many Americans and Europeans became swept up in a bout of spiritualism. Spirit photography was used to capture the flickering souls of the dead, looking like so many fireflies on a summer evening. Alexander Graham Bell thought that the telephone might be used to communicate with the recently deceased. Spiritualists and mediums of every variety frightened both believers and curiosity seekers with elaborate séances as they sought the answers from the other side. The Ouija board (whose name comes from the French and German words for "yes") became a must-have item for the spiritualist set; place your fingers on the planchette, ask a question, and wait until the spirits deigned to answer. When they did, the planchette would sweep across the board, spelling out a reply. It certainly seemed magical, and quite in keeping with the tenor of the time: in the way that electricity was mysterious and invisible, the movements of the planchette seemed to be controlled by secret forces.

Although no one has adequately explained the operation of the Ouija board, current scientific thought holds that the subconscious mind (whatever that is) directs the planchette, using nearly imperceptible movements of the muscles in the arms and fingers, guiding it to deliver the answer. Essentially, the Ouija board becomes a tool for reading one's own mind. Even though this is widely known, many people want to believe that it possesses some sort of power, a connection to an unseen universe of souls who are able to reach us through this simple device. In defiance of all rational explanations, these mysterious spirits from the Great Beyond exert a powerful attraction, and the Ouija board remains popular even today. Witness Ouija 2000.

Ouija 2000 runs on a computer attached to a robot in a laboratory at the University of California. Somewhere in Berkeley, the oracle listens to the disembodied spirits of the Web—my own included—and responds to the gentle caresses of data with sudden movements. Hardly magical, it's nonetheless captivating, an unexpected joining of mysticism and mechanism illuminating the barely explored possibilities of the Web. Far from being a monotonous universe of shops and entertainments, the Web continues to surprise us in its variety, its irrationality; despite all attempts to paint a solid, stable veneer on its surface, it is always capable of transforming into something unexpected.

If others use Ouija 2000 at the same time I'm in the system, we move the planchette together, sometimes in opposition and at other moments in a multiplication of force. It becomes increasingly unpredictable as greater numbers of us try to make something of its oracular gifts, responding less to the movements of my own mouse and more to a synthesis of many minds. But it's silly—very nearly a joke. A Ouija board? Isn't this a ridiculous conjunction of the impressive with the inane? Sitting in the dark, hands gently on my mouse, I certainly feel a bit foolish.

This tension between importance and irrelevance seems to have been designed into Ouija 2000. It's difficult to know if its creators are taking themselves seriously. At once technically sophisticated and meaningless, it somehow manages to bring the Web revolution back down to size, poking fun at our grand pretensions with an antique bit of hype.

It makes me uncomfortable. I am a believer: the Web is the potential ground for an infinity of great works. I can feel this—and sometimes I even get to witness these works. But this twisted, ironic joke deflates my expectations, because I can see my own hyperbole reflected in it, just as the spiritualists saw their own, a century ago.

It has been said that art is a mirror, that it must make you uncomfortable, pushing you out of the zone of your expectations into a new understanding of the world. Strangely enough, Ouija 2000 does summon some powerful forces—the ghosts of every hopeful futurist who brazenly pronounced a new golden age of humanity. It calls them forth from their hiding places, then laughs in their faces. Which must be why the curators of the Whitney Museum of American Art, New York's ground zero for contemporary works, selected Ouija 2000 for their biennial exposition in 2000.

The Whitney Biennial, bowing to the spirit of the time, featured a wide range of pieces employing the Web as a vehicle for artistic expression in works both serious and absurd. (In the case of Ouija 2000, we see equal measures of both.) This brings me back to my first days on the Web, when I noted the lack of artistic content; the biennial promoted the Web as a vehicle for public art, a blessing long overdue. In the case of the creator of Ouija 2000, it also granted public recognition to an individual who might personify a new archetype of the twenty-first century—someone with the mind of a scientist and the soul of an artist.

In these early years of the Web Era, such people are rare indeed, but as human imagination becomes comfortable with the possibilities offered by the playful world, we can expect to see the legitimization of an important aesthetic movement: digital art.

MAKING THINGS GROW

Digital art seems a contradiction in terms to many people. The word *digital* evokes a world of computers and networks and technologies that would seem on the surface to have little to do with the values of beauty and truth that reappear in the history of imaginative forms making up the artistic canon. Yet artists have always been inventive technologists: Da Vinci dreamed up a world of aircraft and war machines, Michelangelo perfected the chemistry of the fresco, and Picasso translated the mathematics of the fourth dimension into the visually confusing shapes of cubism. When photography came along in the mid-nineteenth century, the entire art establishment resisted the high-tech apparatus of silver iodide and long exposures, finding no aesthetic merit in the captured image, until Alfred Stieglitz, in the first decades of the twentieth century, created a series of images so profoundly moving that any argument against their artistic value quickly died away.

The world of digital art lives in a limbo akin to photography before Stieglitz. Though it has been a field of exploration for nearly forty years, no artist has broken through with a work that moves the entire category of digital art into the accepted world of high art. The community of digital artists, dealers, and aficionados, ever conscious of their status as marginal participants in both the culture and business of art, have become ever more insular, defensive, and inaccessible. Once in a great while, however, a work comes along that breaks through the constrictive categories of art to achieve real prominence and popularity.

Every September, Austria's Ars Electronica festival hosts the elite of the digital art world and showcases their best works. Many of these pieces, lofty and theoretical—as if to justify their pretensions to high art—land in the dust bin of history, forgotten almost at once. A very few—the top award winners in the festival's competition—make their way into the Ars Electronica permanent collection in Linz, Austria. Others, snubbed by the festival's judges as uninteresting, still enter the collection because of their unrivaled popularity, including one work that offers a vision of digital culture broad enough to embrace the organic drives of both plants and people.

At the center of the large room housing the collection, a sturdy metal sentinel guards a lush brood of flowers. Arranged in a doughnut-shaped tub encircling a single robotic arm, these plants receive an ideal quantity of artificial light and endless attention from a squadron of victory gardeners—almost none of whom has ever seen the tiny plot in person.

The garden mostly remains perfectly still and quiet, but occasionally, as if it had a will of its own, the metal arm springs to life, whirs and clicks as it changes position and sprinkles water on a tiny shrub. Less frequently it puts a shovel to the earth beneath, digs a hole, and plants a seed. The robot has no visible controls and, to all outside appearances, seems to be operating on its own, almost as if a farmer's soul has been trapped inside awe-inspiring machinery.

Four thousand miles away, in front of my computer's web browser, I have sent that mechanical limb into action with a click of my mouse. The web page I'm looking at controls that gardening robot, and each of my clicks sent around the world becomes another in a series of commands it faithfully executes. As it completes each maneuver, the robot transmits a photograph of its new position—a view from above, peering into the garden—back to my web browser. With each successive click, the robot moves—sometimes an inch, sometimes a meter—responding to my curiosity with a continual array of images that reveals a richness to rival any garden tended by a real person.

This action at a distance, also known as teleoperation, has been a familiar feature of the technological landscape for twenty years, ever since we sent radiation-resistant robots into the dangerously contaminated guts of Three Mile Island and Chernobyl. Looking like simplified versions of R2D2 and equipped with cameras, microphones, probes, and claws, they safely extended our own senses. As our eyes and ears and hands, they allowed nuclear scientists to explore the extent of each catastrophe and then to map strategies for clean-up and recovery. Their close cousins of today, space probes such as Galileo and the Mars Pathfinder, listen to the faint signals of a distant Earth and respond with incredible photography or roving missions to the nearest outcropping of extraterrestrial rock. Robots have become our indispensable mechanical companions, taking us along on journeys we could never make by ourselves.

As a child Ken Goldberg played with the robot hoist at his father's engineering firm and dreamed himself into a world where machines could translate his commands into their actions. Just as I dreamed about a future out among the stars, Goldberg imagined a land where humans and robots worked together to build the world. He grew to adulthood fascinated by the interplay of mechanics and electronics, littered his parents' house with model trains and cars, fired model rockets into the skies, and learned that even the simplest rockets must be teleoperated—it being too dangerous to use them in close quarters. Undergraduate work at the University of Pennsylvania led to graduate studies at Carnegie-Mellon University (perhaps the best school on the planet for the study of robotics) and Dr. Goldberg found himself with the training and resources to pursue his vision. As a newly appointed assistant professor of mechanical engineering at the University of Southern California, he began to cast about for a project that could do justice to his ideas of a graceful meeting of humanity and technology.

In early 1994, Goldberg discovered the Web—"an amazing moment," he recounts—and pondered a marriage of his old passion, robots, with the white-hot Web. Working with colleagues from the University of California, he began to design a complex toy that would bring the still mostly untapped power of the Web into the real world. The basic idea was very simple: robots require some form of human control; the Web provides an attractive, all-pervasive vehicle for human input. Keystrokes and mouse clicks *could* be translated, via another computer, into robotic commands. From this, the Mercury Project was born.

Even if they succeeded in attaching a robot to the Web (a mix of physicality and virtuality never conceived of by Tim Berners-Lee), Goldberg's team found themselves wondering why anyone would stop by, virtually, to move a robotic arm. They needed a reason, a story within which their device could play an integral role. Entertainment, hard to come by on the still-infant Web, provided a platform for their ideas: make it a game, they reasoned, and people will visit our web robot again and again. Given the enormous popularity of entertainment websites today, this seems like a logical chain of thought, but before Comedy Central, *The Onion*, or Shockwave.com went online, this simple idea was a master stroke of vision.

Goldberg started with a semicircular tub of sand about three feet in diameter, in which he buried a treasure trove of objects waiting to be unearthed, examined, and identified. Above the tub he positioned a robot arm holding a high-pressure air nozzle pointed into the sand. That was the basic design, but now came the careful, hard work of testing this device. Would the blast of air be too strong? Would it push the sand out of the tub or gently blow it to one side? How far above the sand should the nozzle be? If that jet of air sent the objects airborne instead, the whole project would fail. So each mechanical detail had to be very precisely tuned to produce an experience that wouldn't quickly end in a messy failure.

Now Goldberg added a cheap digital camera to the arm, providing a photographic view of the sandbox beneath the nozzle. Nozzle and camera, the two indispensable essentials of the Mercury Project, allowed visitors both to excavate objects and to see the results of their labors, a teleoperated hand with a telepresent eye. (Telepresence refers to the ability to monitor from a distance, but it doesn't imply any ability to interact.)

All of the items buried in that tub of sand held clues to a larger mystery; each had been referred to in a work of nineteenth-century fiction. The challenge, as explained on Mercury's web pages, would be to learn enough about the buried treasures to guess which novel Goldberg's team had selected. (The objects were all from Jules Verne's *Journey to the Center of the Earth*, a nice bit of self-reference.) Now they had both a neat toy and cool game, an unbeatable one-two punch that assured them tens of thousands of visits (and repeat visits) at a time when only a few hundred thousand people used the Web with any regularity.

Opening its doors in August 1994, the Mercury Project admitted 50,000 separate visitors over the nine months of its operation, many of whom came back repeatedly. A significant percentage of *everyone* on the Web in late 1994 and early 1995 came to play in Ken Goldberg's sandbox.

Another of Mercury's successes came from its supple social engineering. As no one person had the time or patience to unearth the forty clues waiting beneath the sands, the team provided an area on Mercury's website where visitors could post messages—questions, comments, or proposals—to the entire community of treasure hunters, using a technology akin to a community bulletin board, but with a tracking feature that made it possible to follow the whole thread of long conversations. An electronic chat space, much like the chat rooms on AOL, allowed instantaneous typed exchanges between visitors. They'd swap information about their finds, share data with other explorers, and endlessly speculate on the meaning of all these seemingly unconnected puzzle pieces. A virtual community born around this game thrived from the first moments Mercury went online—a fact not lost on Goldberg. Above and beyond the successful marriage of robot and Web, he'd created a shared place to play, not just a thing to play with.

Using everything he learned, and now partnered with a growing team of researchers, Goldberg began to design the successor to the Mercury Project. One of Goldberg's colleagues, Joseph Santarromana, suggested an organic portal, an environment where the social activity centered around a garden rather than a sea of clues. Given the innovations pioneered in the Mercury Project, it seemed possible that a next-generation web robot could handle all of the physical functions of a gardener, like planting and watering. It would be only slightly more complicated than blowing air into a pile of sand. Moreover, gardening had an appeal far broader than a simple hunt for clues. One of mankind's earliest social activities, gardening could introduce a device like a web robot to a community necessarily blessed with continuity and patience—you can't pull on a plant to make it grow—a community well connected across cultural boundaries. Agriculture is universal.

Soon after Mercury had been broken down, the Telegarden began to rise in its place. Starting with a donated $40,000 robotic arm holding a bristling array of attachments—a shovel, water spout, and hoe—the team added a new full-color camera and dramatically improved the web interface, taking advantage of the latest web browsers released by a newly born Netscape Communications. Now a visitor could click her mouse upon a schematic of the Telegarden—an image showing the current position of the robot arm over the tub of plants—and the robot arm would snap into its new position, precisely where the visitor had clicked on the diagram. You could click on a button labeled "water" and a stream of water would issue from the arm. Another feature allowed visitors to raise or lower the arm in eight increments, zooming in on a single plant or backing out for a wider view of the garden. The camera could transmit a color photograph or it could be programmed to make a tour of the garden, converting a succession of images into a short movie, sending that movie back to the visitor's web browser.

The Telegarden, just over six feet in width, could successfully maintain no more than a few hundred separate plants in its steel-plated tub. More than this and they would crowd each other into extinction. But each of these tiny plots could be independently farmed, monitored and managed. People could come in, choose a plot, and set to work. Goldberg and his team had created a web robot that brought out the full potential of teleoperation—constructive community action at a distance.

The Telegarden premiered at Los Angeles's Interactive Media Festival, the same event that saw the U.S. premiere of T_Vision. Even with this kind of competition, the Telegarden received high praise and the judges' prize. As he and his crew ascended the stage to receive their award, Goldberg found himself full of emotion. His visions of a graceful meeting of human and mechanism had come to fruition. And while Ars Electronica may have overlooked the Telegarden (even as they requested it for their permanent collection), many others saw true genius in Goldberg's work.

The Telegarden caught the eye and imagination of the news media, and was featured on *CBS This Weekend*, and in the pages of *Newsweek*. Goldberg, self-effacing and handsome, became something of a star in interactive circles. The work could have had no better spokesman. Word spread, and for a time Goldberg found himself swamped with media interviews, speaking engagements, and offers to exhibit the Telegarden.

Like wildfire, web surfers around the world pointed their web browsers at the Telegarden website and entered for a stroll through Goldberg's virtual conservatory. By today's hyperactive standards, the interface to the Telegarden seems almost antique, but its simplicity actually reflects the nature of gardening; plants need little help to grow as long as they're given good soil, water, and light. Considering the current trend of overloading web pages with buttons and options and things to mouse around with, the Telegarden doesn't appear to have much to offer a web surfer. But the Telegarden has an abundance of soil, artificial sunlight for some twelve hours a day, and at the click of a button from anywhere in the world, it can be watered.

In the Telegarden, functionality goes hand in hand with community. Anonymous visits to the garden require no credentials, no proof of identity, because guests in the garden can look but not touch. To water the Telegarden or to plant a seed, you must register as a member and obtain a password. More than just a protection against unauthorized (and potentially destructive) activities, membership gives rise to a noisy, vital community of telegardeners.

As the Telegarden nurtured its own community of enthusiasts, the telegardeners began to create an online culture around their shared activities. As with the opening of any frontier, first came the real estate rush: telegardeners selected plots, using maps indicating ground already occupied by other telegardeners. Next, they could press the Plant button on a web page and the robot arm would dutifully dig a hole in the earth, drop in a seed, and backfill its diggings. Now telegardeners would have to sit back and wait and water—and chat. The same chat system that had proven so popular on the Mercury Project became absolutely vital to the Telegarden, serving as the town square for a virtual community built around a real-world project.

Members quickly discovered which plants took to the soil and spread the word. They learned by trial and error how much water each growth required and spread the word. Telegardeners also sized each other up. Friendships formed around shared responsibilities—"you water my garden while I'm on vacation and I'll take care of yours next month"—and the gardeners discovered a shared joy in the process. Like the weather, gardens are a never-ending source of conversation.

The true character of a community becomes visible in its reaction to catastrophe. Unlike the entirely synthetic worlds in the Web, the Telegarden has a physical reality, which means the telegardeners' actions have real-world consequences. The door to catastrophe, always open, admitted its first demon in November of 1995. After a long weekend, the physical curators of the Telegarden returned to find it swamped with water, its tub buckling under the added weight, leaking into shorted-out machinery. Disaster.

The Telegarden had suffered serious damage, with many of its plants drowned in the soup of soil and water. The telegardeners, simultaneously angry and powerless, used the chat system to encourage Goldberg to act quickly and find the guilty party. An examination of the robot's logs seemed to point the finger at a single culprit who had issued the "water" command over ten thousand times (!) in the course of fifty hours. This didn't seem possible; no one could press—or would care to press—a button ten thousand times. Goldberg suspected that someone had written an agent, a computer program that could, under the guise of a member, send an endless stream of "water" commands to the Telegarden. In this, Goldberg recognized his own oversight; the designers had never set an upper limit to how frequently the garden could be watered, a big security hole just waiting to be exploited by a sociopathic computer hacker.

Despite this unfortunate lapse, responsibility still lay on the shoulders of one of the telegardeners. The commands originated with a member of the community, one who had either had his credentials stolen or acted in malice himself.

Goldberg fired off a message, demanding an explanation for the events of the weekend, asking if the member had created an agent that had ruined the Telegarden. The reply surprised him. Unbelievably, it seemed as though someone did indeed press the "water" button ten thousand times. The member in question had no idea that the Telegarden could be overwatered; all of this just came down to an oversight on Goldberg's part that unfortunately had broad repercussions.

The community, satisfied that the crisis had been resolved, began an intense self-examination: could they prevent this from happening again? Should they? Certainly, the Telegarden's designers could add programming that would make it impossible to overwater the garden, and they offered to do so. But, after weeks of debate, the telegardeners decided to live with the danger (and the freedom) of overwatering. (Of course, everyone keeps a closer eye on the Telegarden these days—eternal vigilance being the price of freedom.) Although the potential for a similar catastrophe still exists, so far the community has regulated itself, and no member has ever intentionally tried to destroy the Telegarden, which speaks volumes about the strength of the bonds between telegardeners.

A few months later, a telegardener posted pornographic images on the Telegarden's bulletin board, which had been broadened to include images in addition to text—a disturbing violation of the community. The telegardeners worried about the response of the curators; in the wake of the overwatering incident, they assumed that the bulletin board would be shut down, that they'd lose their vital center in the name of security. But the curators stepped aside and let the telegardeners solve this one for themselves. Everyone already knew who had posted the images, because posts carried the name of the member who made them. That member stepped forward to volunteer his own punishment: a two-week suspension from the Telegarden. He asked the curators that no restrictions be placed on his fellow telegardeners because of his own indiscretions. With the full support of the community, the curators agreed. After that, the bulletin boards stayed clean.

With a fixed amount of land (about a square meter) the Telegarden can never meet all of the agricultural demands that might be placed on it by millions of potential telegardeners, or even its seven thousand registered members. This fact resulted in the only lingering misfortune of the Telegarden—which surprisingly recalls a bit of Colonial American history. The tragedy of the commons—a legacy of the seventeenth-century days when public spaces could be used to graze livestock—reared its nearly forgotten head. When Boston had a population of 3,000, bountiful common lands met the needs of all, but when that figure grew to 50,000, competition for scarce grazing resources resulted in violence. The colonials solved the problem by striking west to the Berkshires, the Hudson Valley, the Ohio, establishing homesteads on land they could farm as they chose. In the Telegarden, while you can farm a plot, you can't *own* it. Any member can water any plot, and any member can plant wherever they like. A telegardener can study the planting maps maintained at the Telegarden website, and, if free space can be found, they can drop a seed into the soil. But with a fixed supply of land, members tend to plant quite thickly, not the ideal condition for the plants themselves.

To solve this problem, Goldberg was forced to implement one last bit of social engineering with an idea he copied from nature: seasons. The concept of seasonal growth followed by a fallow period offered an opportunity to cleanse the Telegarden, bringing it back to its original, uncultivated state. Although the Telegarden resides in a climate-controlled environment, the idea of seasons had an appeal that made it an easy sell to the telegardeners—not that they're ever complacent when they receive warning of the coming winter. Each telegardener feels he or she has put in a special effort bringing the seeds to life, and each sends messages to the Telegarden curators asking them to save his or her own little patch of green. Nevertheless, every six months, the Telegarden is dug up (the plants are relocated outside) and a fresh layer of dirt is laid down. The cycle of land rush, planting, waiting, watering, and flowering can begin again.

You've probably guessed that Ken Goldberg is also the creator of Ouija 2000. The Whitney's curators chose to honor him for that most recent work, but his selection for the biennial also represents recognition of the Mercury Project, the Telegarden, and a number of smaller works. His credentials as a first-rate artist have been established, but Goldberg, first and foremost, is a scientist, and it is in the sciences that his work has its broadest applicability.

The impact of Ken Goldberg's ground-breaking work on web robots has begun to filter back into the disciplines of robotics and communications. His work on the Mercury Project was published in academic journals in 1995, where it stands as the progenitor of a new world where *remote control* means more than just the infrared bursts that cross the chasm between couch and television set; those words now describe how a distant hand, reaching through the Web, can position a camera, lift a shovel, or rain down water.

Because of Goldberg's research, the Web has become even

more than an infinite repository of human knowledge and experience. It has become active, a force in the real world that can place us—not just our eyes, but our hands and teeth and feet and fingers—somewhere else. This kind of puppetry will grow more common as we increasingly see the Web as a magic mirror: we can put our hands all the way through it and shape a world far from our own.

All of this action at a distance makes Goldberg a bit uneasy. The front page of the Telegarden website contains a quote from Voltaire: *"Il faut cultiver notre jardin."* Cultivate your own garden. It's Goldberg's subtle way of telling all of us to walk away from our computers, our web browsers, our telepresent wonders, and explore the world for real.

MINDS AND HANDS

Even if we're not telegardeners, we find ourselves increasingly drawn to the Web. Every day another creation seems to make it more indispensable. But so far, the Web has only offered a pale reflection and distillation of humanity; we are more than a collection of facts, theories, and fantasies. Human beings do things, make things, break things. If we come to learn about the world by playing in the world, the Web must become a playground if it can ever hope to teach us anything about ourselves. And to become a fit ground for human exploration, the Web must assume a more familiar form.

The human body has a wide array of senses, but we consider five to be of supreme importance—sight, hearing, smell, taste, and touch. Beyond these we have the hidden detectors that tell us if we're standing upright or lying down, if our arms are outstretched or at our sides, if our skin feels damp or dry. Each of these sensations tells us that we exist in the world. Take away any one and we are incomplete; lose all of them and we experience a living death. Even the child in the womb is sensitive to light, to touch, to sound, and will respond differently to gentle caresses than to an angry thump. We are wholly involved in a sensual world.

When we spend time in a disembodied universe, such as the mind's space of the Web, we adjust ourselves to it, living more and more inside our own heads while we visit the global mind of facts and figures stored on the planet's array of web servers. The most extreme caricatures of this adaptation are the pale nerds who never see the light of the sun, preferring the gentle glow of the cathode ray tube to more natural radiations. But in a sense we are all living indoors—in our heads—these days, as our eyes, glued to screens, scan the ever-increasing and ever more seductive pages of the Web.

And here the Web presents its greatest danger to us. Like the Pied Piper, it lures us into a world from which we will not even dream of escape. The perfect web surfer is glued to his seat, working a mouse with one hand and a keyboard with the other. His eyes scan and read and absorb, only to scan and read and absorb some more. Although some science fiction writers (and some scientists) have posited a disembodied future for humanity, where our brains, wired into the global networks, silently create a fantasy universe in cyberspace, our existence depends upon our activity. Without action, we lose the basis for our being human.

Although we need to cultivate our own gardens, push back from the screen and take in life in all its reality, we also need to cultivate the Web earnestly. It needs to be equipped with the equivalent of arms and legs and ears and eyes to achieve its full capabilities; without them it will be less than whole—and we will bear the burden of its defects.

Fortunately, we are industrious folks, constantly dreaming up schemes to push the Web into a more substantial reality. Although it may sound a bit silly, the real usefulness of connecting a refrigerator to the Web shows how the insubstantial world of cyberspace and the world of real objects are beginning to intersect and influence one another. The Web helps to make the physical world more versatile, while connections to real objects give the Web a more substantial presence. Vending machines send messages across the Web when they need to be refilled; an automobile can report its location and its driver can receive directions. Each of these tiny steps represent efforts to invest the Web with a material reality, translating it from the ethereal realm of thoughts and facts and grounding it in the very present concerns of the living. This will be *the* major project of the early twenty-first century: the creation of a Web with many eyes and ears and hands, granting it the same senses we have, many times over. And despite Ken Goldberg's ground-breaking work, it is not enough to be able to touch the world at a distance. We need to feel it as well.

In 1989, Atari introduced its enormously popular Hard Drivin' racing video arcade game. Although it differed very little visually from its predecessors, the steering wheel on Road Rash reflected the pull of an automobile. For example, it's far more difficult to pull your car into a turn at 60 miles per hour than at 15. That's because your automobile has inertia; it wants to keep going in the direction it's initially headed. (Power steering makes it somewhat easier, but you can still feel the difference distinctly.) As you cruised around Road Rash's race track, you were forced to make a series of hairpin turns—and the game would push back! This gave the simulation an incredibly realistic feeling. In any of the racing games that had preceded it, you could spin the wheel as much as you liked with ease. Road Rash suddenly introduced a dose of real-world physics—and racing games changed forever.

Amazingly, the inventors of virtual reality systems had overlooked the importance of touch. In the early VR systems of the 1980s, you could don a Data Glove and use it to manipulate objects in the virtual world, but you couldn't feel them. You had to operate very carefully because you weren't getting the all-important feedback from your hands that would have guided you through a similar action in the real world. Those VR systems focused on sight and sound, forcing people to use these senses in place of touch. Yet the sense of touch, immediate and instinctive, is often the best way to move through a foreign world. This oversight (no doubt due to the fact that most VR researchers had begun their careers in computer graphics) probably contributed to the failure of many early VR projects. Seeing might be believing, but touch is undeniably real.

In 1998, when Sony marketed the Dual Shock Analog Controller for its PlayStation, home video gamers were introduced to the concept of haptic (touch) feedback. Computers can draw complex pictures at breathtaking speeds and play entire symphonies from synthesized sound, but until a few years ago, they were unable to reflect our tactile sense. Human beings possess a highly developed sense of touch; our fingertips are crowded with nerve endings that can instantaneously identify smooth, rough, hard, soft, wet and dry, hot and cold. We're also gifted with nerves in our joints and muscles, so we can sense how much force we're applying to an object within our grasp. All of this keeps us from hurting ourselves; without this haptic feedback, we'd likely sprain our wrist every time we tried to open a mayonnaise jar.

As our reach extends beyond our bodies, using robots and distant sensors, our need to touch the world around us and to be touched by it becomes ever more important. To create a real feeling of presence at a distance, telepresence, we have to be able to feel our way through the world. It's the next best thing to being there. Although Atari and Sony began to explore haptic interfaces, these toys were only primitive preludes to amazing new innovations in the feel of the virtual world.

In 1997, as I toured the trade show floor at SIGGRAPH, I came across the modest booth of SensAble technologies. They were showing something that looked like a drawing pen attached to a movable arm. This seemed to be connected to a computer, whose screen displayed a few child's blocks.

"What's this?"

"Haptic feedback. Give it a try."

I put the pen in my hand and immediately realized that I could drag one of the blocks around on the screen by moving the pen about. Not very interesting. However, when I tried to move that block over one of the other blocks on the screen, the pen came to a screeching halt as the blocks touched. It wouldn't move. It felt as if the blocks were solid.

Someone typed something on the computer's keyboard, and suddenly the blocks were—squishy! I could ram the blocks together using the pen, and they'd give—just a little bit. It felt as if they were rubber, capable of being deformed up to a point. I was blown away. I'd read about the haptic interfaces, but I'd never used one this sophisticated. Here was a real haptic interface, connected to a computer simulation, and I could buy it today! "How much?"

"Around ten thousand dollars."

Ah well. My dreams of owning one of these sweet toys were gone—for the moment. But the following year, SensAble introduced a scaled-down version of their product, which cost little more than a thousand dollars. As we become more and more familiar with haptic interfaces and understand how important they are to us, we'll see them plummet in price and become nearly as common as those ubiquitous mice attached to our computers. Then, we'll be able to touch the world at a distance and feel it push back.

PANOPTICON

Despite Voltaire's warnings, I'm sitting in front of my web browser again—but this time I'm looking out onto a night scene, far above San Francisco's waterfront. A camera perched at the edge of downtown's Embarcadero Center points out into the darkness; a string of lights illuminates that city's Bay Bridge and the numerous wharves that dot the shoreline. The picture is a bit blurry—it can't be a very good camera—but that's what San Francisco looks like *right now*. My eyes can be somewhere far away—telepresent, as Ken Goldberg would have it—while my body remains in front of my computer.

This live image (it updates every few seconds) comes to me courtesy of one of the most intriguing sites on the Web, Earth-Cam. The site, which looks very similar to Yahoo!, contains a fairly exhaustive list of all of the cameras attached to the Web—webcams, as they're called. The webcam itself began as a prank. Some engineers at Netscape pointed a camera at an aquarium in their offices, had it snap a picture every few seconds, then wrote a computer program to convert the image to something that could be viewed within a conventional web page. The page would then update automatically every time the camera snapped a new picture. All you had to do was point your web browser at the page, and you'd get a series of images from the tank, like a time-lapse movie.

Soon cameras grew more numerous, and a few even presented real-time, TV-quality video, if you had the huge amounts of bandwidth needed to receive thirty images a second. Some looked out on pastoral scenes, others pointed into college dorm rooms in various states of disrepair. Others allowed web surfers to voyeuristically peek into the lives of tele-exhibitionists. One young woman named Jennifer decided that the webcam represented an entrepreneurial opportunity and set up the Jenni-Cam, a series of cameras that captured her entire Washington D.C. apartment, including her bedroom, allowing guests in for peeks. For a fee, you can have nearly unlimited access to the minutiae of Jennifer's life, watch her sleeping, reading—or, on occasion, making love. Not exactly reality TV, not exactly pornography, the JenniCam offers visitors a look into someone else's life—for as long as they might care to peep. (With appearances on shows like *Late Night with David Letterman*, Jennifer has become something of a celebrity.)

There are now hundreds of thousands of cameras attached to computers, and most of these computers are on the Web; each of these, at least in potential, can become a webcam. EarthCam is attempting to index them, much as Yahoo! has indexed the content of the Web: by region, by subject, by use. You can open up your browser and check the traffic before you take the morning commute, or peek in on the space shuttle, or just watch Big Ben as its hands trace out the hours of the day. Your eyes can be nearly anywhere around the globe the Web has touched. You can linger or hurry on to the next scenic locale.

If the Mars Polar Lander had not crashed into a canyon on Mars's South Pole, we would have been granted an interesting extraterrestrial experience. The probe was equipped with a microphone, which would have beamed the sounds of the Red Planet back to an eager audience on Earth, ready to hear something that would have been literally out of this world. On the day that Mars Pathfinder sent its first images back—July 4, 1996—NASA's servers were flooded with over *100 million* requests for the pictures. Although not real-time (the photos were a few hours old by the time they made it onto the Web) they still revealed the ability of the Web to propel us into other worlds, to take us by the eyes and send us on a journey.

Today we cultivate the Earth with the Web: Ken Goldberg has taught us how. Tomorrow, perhaps, we will touch other planets, reaching out through the Web to get a feel for things we have never known before. We may prefer to cultivate our own gardens with our two hands, but there are so many places we cannot reach by ourselves. The Web has become like a spaceship, carrying our senses to the edges of the known. The Web can trap us in our own illusions, or it can show us things we have never imagined. It can even give us a new sense of the earth beneath our feet.

THE GREEN GODDESS

In the early years of the twentieth century, R. Buckminster Fuller contemplated the disaster of his life. The Harvard graduate had failed in his business, and he had a wife and a newly born daughter to care for. Deep in a depression which seemed to have no resolution, Fuller stood on the shores of Lake Michigan, contemplating suicide. Something deep within him resisted the idea, and suggested an alternative: what if he just listened to everyone and everything around him? Perhaps if he listened closely enough, he'd find out what he should be doing with his life. He knew for sure that his own attempts to create a life had brought him to complete failure. Anything, however outlandish, would be better than his present situation.

Fuller didn't speak to anyone for an entire year. Not to his wife (who regarded her husband as mildly insane) or to his friends. Instead, he became a keen observer of the world and thought long and hard about what was worth doing. In 1928, his silence ended and he set to work. Over the course of the next fifty-five years, Fuller completely revolutionized every field he touched.

Fuller had a degree in naval architecture, but what really interested him was design, all aspects of how human artifacts came to be made and used. By the 1920s, the industrial economy was in full swing. Mass production of the automobile had crowded the streets with cars; radios allowed people to learn of events far away from their own homes; medicine seemed to promise a longer life span for anyone wealthy enough to get quality care. Fuller realized that these seemingly distinct trends were really part of a wave of prosperity shaping the human race into the most successful species in the planet's history, and that there was no end in sight.

From his study of history, Fuller knew that living standards for even society's poorest had improved tremendously during the Industrial Revolution. Average members of the middle class had a living standard that equaled or exceeded that of a sovereign during the Middle Ages. Everyone was moving up the economic scale, though not at the same rate. Fuller charted out this growth in wealth and calculated that, by the end of the century, the average person would have the same living standard as a millionaire of 1928.

How could this be possible? Fuller understood the real nature of technological innovations; each invention helps humanity to do more with less. As we learn more and more about nature, about the physical world, we can use it more effectively, achieving the same goals with less effort and fewer materials. This suggested to Fuller that the same resources could be spread ever thinner, until everyone's needs had been comprehensively met.

In later years Fuller wrote:

> For the first time in history it is now possible to take care of everybody at a higher standard of living than ever known. Only ten years ago the "more with less" technology reached the point where this could be done. All humanity now has the option to become enduringly successful.

Although many of his contemporaries considered Fuller a wide-eyed idealist, he was nothing of the sort. He saw the economic and political realities of his day and accepted them. Nevertheless, he argued that the ever-more-evident technological realities of doing "more with less" would eventually triumph, ensuring a successful future for all.

Fuller did more than preach his gospel; for over half a century he became what some called "the planet's friendly genius," coming up with a stream of ideas that would prove his point beyond any dispute. If we wanted a future with success for all—and Fuller decided this was a quest suitable for his life's work—we'd need to learn how to build for global success, pushing technology to its limits in the quest for efficiency, economy, and design elegance. He promoted what he called a comprehensive design science revolution, which would utterly remake the way we construct the artifacts of civilization.

Several of Fuller's inventions have become important features of modern life, but none more so than his geodesic structures. Based upon his own system of geometry—perhaps the first real innovation in geometry since the ancient Greeks—Fuller designed structures that reinforced their own strength in their essential design. Unlike conventional buildings, which had weak points that would invariably lead to failure and therefore had to be expensively reinforced, Fuller's geodesic structures spread their loads across their entire volume. No single point could fail in a geodesic structure, because the entire structure shared the entire load equally. This meant that structures could be built out of a hundredth of the material that had previously been used, an economy which, Fuller hoped, would lead to an era when everyone on Earth was housed in a decent, sturdy dwelling.

Today, temporary and permanent shelters of all types are constructed from geodesic principles. Modern tents used by campers are built from light aluminum frames arranged in geodesic patterns. Montreal's Expo '67 greeted its visitors with the world's largest geodesic dome, a gigantic sphere twenty stories high, an enormous construction that used practically no materials. (The curious paradox of geodesic structures is that as they grow larger, they both become stronger and require fewer materials per cubic foot of volume.) Today, we buy our children playground sets formed in the easily recognizable half dome of a geodesic sphere. They're inexpensive and safe. Fuller would be pleased.

The Geoscope, one of Fuller's most ambitious projects, was never fully realized in his own lifetime. In the 1950s, Fuller had calculated that a geodesic sphere 200 feet across would have enough surface area that, if photographs were laid upon it, you'd be able to make out features of the Earth's surface, down to individual homes. In a lecture he gave in his closing years, summarizing his life's work, he talked about the Geoscope:

> . . . if you took the 35mm photographs [of the Earth] made by aerial photography, and you put these 35mm photographs together, edge to edge, it would make a 200-foot sphere. You could take a direct photograph. And in those photographs, you could make out, you could see all the streets and everything, you could see individual human houses, but you can't see the humans. But you can see your home. You know that's your home just as clear as can be, you can pick it out. So I wanted some way in which you had a scale where human beings could really feel themselves on the Earth, even though they couldn't quite see themselves, they could really feel, these are my works, and the house is part of me, so that was the scale.

Even before the invention of the satellite, Fuller knew that it was already possible to represent the entire Earth photographically. Today, weather satellites and spy satellites and the space shuttles give complete and continuous coverage of the planet, in eye-popping detail. (It's known that spy satellites can read the license plates on automobiles, and make out the rivets on airplanes.)

Fuller proposed constructing his Geoscope over the East River of Manhattan, next to the United Nations, so that the world's leaders would be able to look out on a model of the Earth and understand—viscerally and immediately—their fundamental concern: the well-being of the planet and its peoples.

ART+COM's T_Vision is Fuller's Geoscope realized. Using computer simulation, T_Vision provides an even greater level of detail than the Geoscope. More than just providing a view of individual buildings, T_Vision can take a look inside them, moving between global and human scales seamlessly. Fuller likely would have been delighted with T_Vision and doubtless would have dreamed up a hundred good uses for it.

A broadly networked world can offer more than just a static view of planet Earth. Every day, the number of webcams is increasing, more portholes into reality are opening up through cyberspace. A few webcams have already been integrated into T_Vision, connecting the actual to the virtual. It is technically possible to integrate the thousands of webcams into T_Vision; it would then become possible to travel to nearly any populous corner of the earth, and look upon it, both virtually and in reality.

Why would we want to engage in such extravagant navel-gazing? Once again, we can turn to Fuller. He realized that the Geoscope would help human beings to grasp not only the size of the Earth, but its complexity. He believed that a comprehensive change in human behavior could come only from a broad understanding of how human civilization works, and that could come only from an experience of humanity at its most comprehensive level. In the 1960s Fuller created the World Game, a simulation of the wealth and economies of nations. Players adopted different nations, then acted as the "stewards" of their national entities. The players would learn how to cooperate (and compete) in the struggle for resources, for power, for advancement. Fuller hoped that this bird's-eye view of the large-scale systems of Earth would teach people that cooperation is *always* more profitable than competition, that sharing inevitably leads to abundance, while restriction and selfishness lead to failure and eventual destruction.

In his attempts to make the World Game as comprehensive as possible, Fuller continued to add every bit of data he could get to his simulation. Before computers became widespread and inexpensive, this mountain of facts couldn't be conveniently processed. Too much information was interdependent, and without any way to manage that complexity, the World Game couldn't be much more than a toy, dealing with the gross relations between nations, lacking the real subtleties that underlie local and international economies. Fuller died just as the personal computer was becoming a common instrument, but something does exist that suggests what Fuller's World Game might be like if he had put it into electronic terms.

When Will Wright wrote the first version of SimCity, he created a complex system of rules that governed a local economy and culture. In the years that have followed, SimCity has grown more complex, more realistic, better able to create a faithful simulation of the real world. It has become an important tool to teach people the relationships between economics, development, and human behavior. Could it be that we are ready to take the ultimate step, to create a simulation within which we could recreate the whole world?

Such an idea is hardly outside the bounds of possibility. Most multinational corporations run sophisticated simulations that help them to predict the behaviors of international markets. Although sometimes inaccurate, these programs help them to chart the uncertain waters of the future. Shouldn't something like this be broadly and freely available to help us all to understand the richness of the world, to aid us in our collective decision making?

Already there are signs that this is happening. I can surf the website of the Environmental Defense Fund's Scorecard, type in my zip code, and find out about the environmental dangers in my neighborhood or in any neighborhood where I might be planning to move. I can learn about income distributions, courtesy of the U.S. Census Bureau. I can hop into Microsoft's TerraServer and get a high-altitude, high-resolution picture of my own home. The information is all there; we need a framework to sew the pieces together, and computers powerful enough to make it all seem perfectly clear, even to a child. The World Wide Web is the foundation for that framework, and Sony's PlayStation 2 would serve well as a supercomputer that could aid us in making sense of the world.

If the twenty-first century is to be a global century, we need to understand how our own decisions affect our neighbors near and far, and we need to understand how their decisions affect us. No one is an island—no one ever was—but now we can make that perfectly clear.

We can do more than watch at a distance, silently observing the actions of others. We can act ourselves. Ken Goldberg's Telegarden has convincingly shown that it is possible for us to take the eyes offered to us by cyberspace, and, together with distant, mechanical hands, move the material world, no matter how far away. We will cultivate our own garden, but that garden is no less than the entire planet.

Some people will no doubt find this idea terrifying. It implies a future where we can be watched, continuously, by a billion unseen eyes. On the other hand, it also means that we can count ourselves among the watchers. There will be no Big Brother, no panoptic central authority dictating the future for all humanity. As we learned through the history of democracy, we will watch, we will argue, and we will act. Together. We will hang together—because the alternative is nearly unthinkable.

This does not spell the end of privacy, but it does set the limits of private behavior; your home will still be your castle. (We should regard with suspicion anyone who suggests otherwise.) Outside of its walls, you may be observed. But even the nosiest of us hasn't the time or temerity to watch the infinitely boring details of the lives of others. Some pathological individuals will no doubt resort to cyberstalking, watching the famous (and infamous) in every second of their public lives. Most of the rest of us will carry on much as before, but with a far more comprehensive awareness of the world around us: the *entire* world around us, in all of its richness, its depth, and its wonder. We will make more informed decisions, accounting for the comprehensive repercussions of our actions, or we will soon learn if these acts have caused harm to another. The closer humanity grows, the less room there is for unconstrained human activity. While this doesn't spell the death of freedom, it does mean that we will need to think through our actions carefully. Anything else will be seen as thoughtless, rude, or simply unfriendly.

Of all the gifts we can give our children to navigate the course of the twenty-first century, a comprehensive model of our Earth must rank among the most important. We need to replace the teacher's globe, a fixture in every classroom, with a living model of the Earth. In T_Vision, one becomes immediately aware that political boundaries are, for the most part, entirely arbitrary, that the governments dividing people are figments of our cultural imaginations and have very little to do with the actual realities of life on this planet. Our children should understand that the political residue of the twentieth century need not constrain their own thinking. Globally, they can plot a course to success which will reach into every corner of the Earth. It is possible, and, if Buckminster Fuller is to be believed, inevitable. He prophesied a critical point in human affairs, when humanity would make the "decision for success—or for extinction," and set 1989 as the date for this collective decision-making process. As we glide into the twenty-first century on a wave of extraordinary potential, I would suggest that humanity choose success, for everyone.

This success comes with a price; every gift of technology can be used for good or for ill. We must take great care to ensure that the future before us becomes more than the amplified echo of the past. Fortunately, our children, equipped with the revolutionary technologies of planetary awareness and global action, will know how to make this a reality.

Our child, on her eighteenth birthday, becomes a citizen of the republic, with full rights and full responsibilities. Both of these she understands in equal measure, having prepared for this day over the last several years. Already she's spent months in front of her PlayStation 5 looking at the Earth. She began by studying it from a distance, watching the slow swirl of clouds across the oceans and continents. Eventually she tired of the play of sun and water and turned her attentions closer to home.

Now she brings her view down, to her nation, her state, her city, and finally her own community, dialing into cameras here and there. It looks like the town she knows, the town she's known from her earliest days. But, in this simulation, she can discover and explore all of the complex relationships underlying the physical reality of the community. How much energy do these homes use? Or these cars? Where does this trash go? Why is the factory down the street emitting such a cloud of smoke? What's in that factory? What do they make? Who buys it? Why?

She begins to understand the tentative web of economics and culture that comprises her own home, then expands her view to the entirety of her town. She sees for herself that most of the rich people live in one neighborhood, while poor people live in another. She wonders about the factors conspiring to create these communities and begins to draw her own conclusions.

Now she broadens her scope, looks at the county, with its farm belts outside of city limits, looks at their land use—and the constant erosion, dirt silting up the rivers, flowing downstream to the ocean. She notes the differences between the organic farms, which use no pesticides or fertilizers, and the conventional farms, which have come to rely on an ever-increasing supply of stimulants to raise nature's bounty. She wonders how long that situation can persist and so searches the Web for analogous cases. She finds one years ago, in a state far away, yet uncannily similar. A situation that ended in desertification. An ecological disaster now waiting to happen in her own backyard.

Before she received her citizenship, she could only watch and learn. Now she can act. She sends messages to a few thousand people. Some are friends, some she's never met. She outlines the problem, pointing them toward all of the resources she's uncovered to help her to understand its seriousness. Soon enough, she receives a deluge of replies. Some people agree, some disagree, some simply don't want to be bothered. A noisy discussion ensues and positions are taken. Some people refuse to believe her data and offer counterexamples, which she studies with the same intensity she gave to the data supporting her arguments. There is some truth in them, and she modifies her position, just a bit, hoping to come to an accommodation. Again, some remain unswayed, but a consensus is building. In the midst of all this discussion and debate, a road to resolution begins to appear. Now she proposes a vote.

The ballot is open to everyone who might be affected by the ruling. This is a significant number of people, some of whom live far beyond the affected region but will still be impacted economically. They too have voices in the decision making. She presents her arguments, the consensus of the citizens she has rallied under her banner, and waits for the results.

With the ballots counted, she finds out that she has won the day. A recommendation is sent to county officials that the land-use policies for these conventional farms be adjusted so as to be friendlier to the environment. As champion of the initiative, it is up to her to monitor the situation and see that these changes are carried out. In her own small way, she has helped to heal a corner of the Earth.

CONCRESCENCE

In the early years of the twenty-first century, the typical home entertainment system will include devices like Sony's PlayStation 2, capable of bringing incredible virtual reality worlds through the Internet into the home.

Computer simulation is on a fast track toward the creation of a new generation of products that will make the virtual seem virtually real. In November 1999, Sony's vice president of research and development, Shin-Ichi Okamoto, reported that he felt that Sony had failed the game development community with the PlayStation 2—only 300 times faster than the PlayStation, and much less than what they'd really wanted. The next generation, he announced, the PlayStation 3, slated for introduction in 2005, would be *1,000* times faster than the PlayStation 2.

If Okamoto can keep his promise (Sony's track record now looks pretty good) the PlayStation 3 will be even better than the real thing. The images it generates will be completely indistinguishable from the real world, as complex and as rich as anything you might encounter in real life.

In the five years between the launch of the PlayStation 2 and the introduction of the PlayStation 3, we'll all be witnessing the power of computer graphics to bring virtual worlds to life. After 2005, the simulated and the real will look pretty much the same, especially when viewed in the unprecedented clarity of high-definition television. It might be CNN, or it might be a simulation of CNN. Like the old Memorex ads, we might wonder, "Is it reality—or PlayStation 3?"

What can we do with this unprecedented power? Play games? Yes, that's a given. We can entertain ourselves endlessly. But we will likely move far beyond that. The "make pretend" worlds of computer simulation bring us opportunities that reality doesn't easily offer. We are at the threshold of a revolution in human experience. Computer simulation is starting to be used as an engine of the imagination, bringing to light some of the most intangible aspects of our being. Just as music, dance, theater, and poetry have helped us articulate the quiet parts of ourselves, simulation will become a new aesthetic, an art form with its own power to illuminate the depths of our being.

There are already examples suggesting what is to come, new worlds that exist only in the circuits of powerful computers, yet are capable of opening our minds and altering our perceptions. When we change the way we see and touch the world, we change ourselves.

CATHEDRALS OF LIGHT

In the mid-1980s, Char Davies set aside her successful career as a painter to become a founding director of a tiny start-up computer graphics company in Montreal called Softimage. Its product, which allowed artists to create 3-D computer graphics without a deep knowledge of mathematics and computer programming, was a big hit. Used to create the special effects for such blockbuster films as *Terminator 2, Independence Day,* and *Armageddon*, it soon became a must-have tool. Softimage's staff swelled from 7 to 200 people in just a few years, and Davies, who had worked nonstop, finally took a few weeks off.

On the sparsely populated Bahamian island of Andros, Davies learned to scuba dive. Over several visits she mastered the technique of deep dives into the Caribbean waters—to the edge of safety, over 200 feet into a 6,000 foot deep abyss.

These dives introduced Davies to a whole new bodily experience of space. Scuba had allowed her to feel suspended in space; she gloried in the enveloping environment of the warm sea. All around her, in every dimension, the water supported her journey. She could move in any direction, free from the tyranny of gravity. She began to consider how she could portray this experience of rapture in an artwork.

Throughout her life as an artist, Davies had wanted to evoke the sensation of enveloping space. It was this desire that propelled her to learn to dive, and it also drove her to master three-dimensional computer graphics as a technique for breaking through the two-dimensional frame of paintings. Beginning with a series of three-dimensional still images, known as the Interior Body series, she took the aesthetics she had already developed as a painter, and using transparency and translucency, created digital imagery that looked startlingly real. One composition, titled *The Yearning*, appears to be a coral ice form captured beneath a quiet sunlit sea, more evocative than any photograph and absolutely alive. Now Davies wanted to go beyond the static image and translate the feel of her scuba experience of enveloping space into a virtual world.

By the middle of the 1990s, computing power had become inexpensive enough that Silicon Graphics could build enormously powerful machines devoted to computer simulations, including their Reality Engine, which began to touch upon that fabled figure of 80 million polygons a second. Their marketing folks imagined that such a product would be used by scientists and engineers, and perhaps even for entertainment. It had the kind of power Davies would need to express her aesthetic, a translucent world that would portray the sense of freedom she felt beneath the sea.

During these years, Davies sought company in the writing of philosophers. *The Poetics of Space*, written by Gaston Bachelard, was particularly inspirational:

> By changing space, by leaving the space of one's usual sensibilities, one enters into communication with a space that is psychically innovating. For we do not change place, we change our nature.

For Davies, this meant if she could *change* space, create a new place for her art to be experienced, she could change the nature of those who came to view it. To do this, she would need to immerse the viewer within her work so that the distinctions between art and observer would collapse. No longer would her images hang in a frame, static and isolated. Now people would enter *into* her art and become part of the world she created.

She knew it was possible, technically, to use virtual reality to achieve such an end, although no one had yet created a virtual artwork of such magnitude. The machines were capable—but could the creator rise to the challenge? How could Davies redefine space to produce a new human nature?

One solution to this, as it happens, was found long ago; you need only walk into a cathedral to understand. In May 1996, I went to Paris for the Fifth International Conference on the World Wide Web. I gave a morning lecture on VRML fundamentals, then struck out to explore the city. In the early evening, I found myself on the Ile de la Cité, a tiny island in the Seine that is the heart of Paris, standing before Notre-Dame de Paris, the world-famous cathedral. The chaotic streets around Notre-Dame are thronged with tourists, merchants, and local residents. As I crossed the threshold into Notre-Dame, a sudden and unexpected sensation enveloped me. The somber quiet of the cathedral space, soaring upward to the heavens, produced a similar sense of peace within me. In a journey of just a few feet, I found myself, my being, in an entirely new space. It was as remarkable as it was unexpected—but this surely was the intent of the artisans, who, nearly a thousand years ago, labored to build this monument to God's glory. Bachelard, as a Frenchman, understood the power of space to transform human nature; he had a defining example at hand.

Instead of a cathedral, the warm, boundless embrace of the sea had transformed Davies. To help her recapture this feeling, she had the ultimate tool for the transformation of space, a Reality Engine capable of extraordinary feats of computer simulation. If she wanted to dissolve the boundaries of being, she could do it by designing a space that contained no distinct boundaries between self and world. This was very much against the tenor of virtual reality, which normally featured sharp-edged objects floating in a black void, looking like an amusement park of the mind's eye. This is what Davies wanted to attack with a direct assault on two fronts: by changing what the eye would see, and by changing how the participant would move through the world.

Davies began to craft a virtual environment of incredible translucency and ambiguity. The artwork would have visible forms, but each would be semi-transparent; you could look through anything and see what lay beyond it. In this work, everything would always be visible everywhere. She created a world of natural-looking forms centered around a great old tree barren of leaves. The tree was surrounded by a forest grove, a pond—with an abyss reminiscent of Davies's scuba experiences—and underneath it all, a subterranean world of rocks and roots.

Then Davies began to focus on the other major element of her artwork: navigation. How best should the body move through this space without boundaries? Davies loathed the computer joystick. It felt too much like controlling a machine. Why couldn't people float through the work, just as she did on her dives? In the underwater world, position is controlled by buoyancy. The human body floats in water, but weights attached to the diver tip the balance, and the body glides downward in a gentle descent. This is so carefully managed that by changing the amount of air in the lungs, a diver can rise or fall. Fill the lungs with air, and the diver rises. Exhale, and the diver falls. It seemed perfectly natural to Davies—breath and balance.

The navigation interface to her artwork managed to reproduce this effect perfectly. Surprisingly simple, it was composed of components costing just a few dollars, including some springs, a strap, and a variable resistor. The springs, when placed in a band around the chest, could change the value of the resistor, depending upon the extension of the chest. Breathe in, extend the chest, and the computer would sense it. Exhale, compress the chest, and the computer would sense that. Suddenly the computer could respond to breathing. Now Davies could re-create the freedom of movement of her underwater journeys.

Months of eighteen-hour days preceded the premiere. In the last months, a team of two audio artists joined her existing crew (Davies worked with John Harrison, a programmer, and Georges Mauro, a computer animator) to add an all-important element—the sonic environment. Davies wanted something that wasn't so much musical as it was evocative. The duo sampled a male and a female voice, then wrote programs that would interactively generate an ever-changing sonic environment that sounded both elegiac and organic.

On August 19, 1995, the work premiered at Montreal's Museum of Contemporary Art. Davies's name for the work said it all: *Osmose*. The French word for "flows between"—and the root of the English word *osmosis*—*Osmose* offered individuals a chance to enter another space and experience another mode of being.

The installation was entirely immersive. Visitors to *Osmose* would don a head-mounted display unit and the chest-measuring apparatus, then the simulation would begin. (While head-mounted displays may be dangerous in the long term, a short exposure is relatively safe.) Inside *Osmose*, people would find themselves apparently weightless, floating in a translucent world of natural forms. Float to the central tree, into its bare boughs, and you'd fade into a lush garden of leaves. Float into a single leaf and you'd be surrounded by green, with tiny bubbles of white "life energy" flowing all around you. Exhale, let yourself drift down, and you'd find yourself looking up at the roots of the tree, surrounded by semi-transparent rocks.

The intense intimacy of *Osmose*—one person, alone within the world—created a challenge for Davies. Although it had always been her primary intent to provide a solitary, contemplative experience, the museum's curators wanted to be able to show the work to a wide selection of people, including those who wouldn't be voyaging through it. To answer these problems, Davies designed an installation space featuring a projection screen so that others could look into a voyage as it was happening. In addition, the voyager—or *immersant*, as Davies calls them—would be standing behind a translucent screen, gently and warmly illuminated. This created a "shadow box" effect; you could watch a person as they moved through *Osmose*, and you could gaze upon what they saw and hear what they heard. Although very intimate, *Osmose* was simultaneously a public experience.

In early August, as she tested the work on a few colleagues, Davies discovered another effect of her transfiguration of space; when space changes, so does your perception of time. Her test subjects would go into the work and come out forty-five minutes later (the limit of comfort for the heavy and confining head-mounted display), thinking they'd only been inside the work for ten or fifteen minutes. She had planned to allow the museum to usher people through in twenty minute intervals; this enabled only about thirty people a day to experience immersion in *Osmose*. Davies realized that people would never leave the work on their own in such a short span of time—it might only seem like five minutes to them. Throughout the development of *Osmose*, Davies had considered various effective "endings" for the work, but only when *Osmose* was nearly complete did she envisage an ending that seemed to flow naturally from the work itself. It was necessary not only to bring people out of the work, but to do so in a way that was very gentle. After fifteen minutes in *Osmose*, the immersant would be gently lifted above the central tree and see it become enveloped within a translucent crystalline form. Then he or she would float away as the world faded to black emptiness.

Montreal artist Henry See was among the first to immerse himself in the completed work at its premiere. When he emerged from his journey, he was silently crying. "It's just . . . very beautiful," he said, then went silent again. (He later told Davies that he had found a quiet corner of the museum and wept for fifteen minutes.) This was not an unusual reaction. Davies kept a log, where participants could record their feelings after their journey through *Osmose*:

> "I always knew, but now I have proof—I am an angel!"

> "Floating. Gently falling. Breathing. Exploring. In delight, the wonders of a green universe. Merging within another creation, but no fear, instead, breathe in, inhale a world."

To immersants within *Osmose*, the ending seemed to be a natural, final act, a sort of near death experience, as they felt themselves drawn up and away from the fleeting beauty of the world. One excitedly wrote in the log that she was "no longer afraid of death." In the changed space of *Osmose*, like some modern cathedral of light, one could find reverence, wonder, and peace.

Osmose was immediately and broadly hailed as a breakthrough work, and touches on themes recurring throughout all of Davies's art: the natural world, raptures of the deep, longing and loss. In late October 1995, crowds lined up in Soho on a chilly evening for its New York premiere, just to catch a glimpse of the piece which had redefined the possibilities of virtual reality. In Montreal, London, Mexico, and New York, *Osmose* was always fully booked with visitors, people ready to take a journey to another space.

Although many museums around the world have been intrigued by *Osmose*, very few are technically equipped to handle the installation, which requires a half-million-dollar computer, a full-time guide, and a small fleet of technicians to install the work. If that's not enough, the limit on the number of visitors the work can receive is a final barrier. Museums, built in the age of art before the interactive era, can appreciate the marvel of *Osmose*, but they can't or won't pay the price to display it. The ten thousand immersants who have experienced *Osmose* represent only a tiny fraction of those who would willingly don a head-mounted display and voyage into Davies's world.

But *Osmose* was created several years ago, on large computers that have less power than those selling today at the local toy store—a development that intrigues Davies. She has begun to wonder if Sony's PlayStation 2 might not be an ideal platform for the presentation of her future works. Sony already sells a device, known as the Glasstron, which provides a head-mounted display that connects to the PlayStation 2. (The potential side effects of this device are unknown to the author.) The breath and balance interface needed for *Osmose* cost just a few dollars to build, so it could certainly be provided to a mass market of consumers hungering for an experience of space that could transform being.

In a few years' time, maybe we will all have access to Davies's cathedrals of light, visiting them as we touch different sides of ourselves. Just as the medieval residents of Paris used Notre Dame, so these new artworks can help us locate the quiet spaces within each of us.

TUNNEL OF LOVE

In 1981, as Eric Drexler worked at MIT developing the foundations of nanotechnology, two Swiss scientists stood on the verge of inventing something that would prove invaluable to the generations of nanotechnologists to come. At IBM's Zurich Research Laboratory, Gerd Binning and Heinrich Roher were putting the finishing touches on a device so simple and yet so profound that they would be awarded the Nobel Prize for their work just five years later.

What was it that garnered such attention and praise? A sharp pin. Using tungsten—one of the strongest of the naturally occurring elements—honed to a very fine point, the researchers fastened their pin to an apparatus that placed its tip just barely above a surface of gold foil. Nothing prize winning in that. But when they ran a very small electrical charge through their tungsten pin, something very interesting happened.

Electrical charge is carried by electrons, which are almost infinitely small subatomic particles. Every atom has electrons circling around its nucleus, moving so quickly that they form a cloud of electrical charge. Electrons have the curious property of being both tiny points of charge and waves of electricity; because they're so very small, they demonstrate both of these qualities simultaneously. In metals, which can transfer electric charge (like the wires in your home), the electrons can move, almost as if they were a fluid, from atom to atom. In a ceramic dinner plate, the electrons are chained to their atoms, so ceramics can be used to insulate against electric charges. (Think of a microwave oven. You can't put metal objects inside of it, because the microwaves will send those electrons shooting about and sparking. But a Pyrex casserole dish—glass is a ceramic—will do just fine.)

Gold is among the very best of all the metals at conducting electricity and it's used in many computer circuits for just this reason. Electrons in gold foil practice a subatomic form of free love, exchanging atomic partners at will. And because of the principles of quantum physics, which state that *everything* in the universe can be considered as both a particle of matter and a wave of energy, electrons have the curious ability to teleport themselves across short distances and enter *into* solid matter, something that shouldn't be permitted. After all, solids are solid precisely because they don't admit visitors—except electrons. This effect is known as tunneling.

Binning and Roher thought that perhaps with their sharp pin they'd be able to detect the tunneling of electrons into the gold foil. (Gold, because of its free-love electrons, can be rolled into foils that are only a few tens of atoms deep, very thin indeed.) Since the pin carried a weak electrical charge, they believed that electrons would tend to leap off the pin and tunnel into the gold foil. When this tunneling occurred, they thought they'd be able to sense it, because the number of electrons passing through the tip of the pin—measurable as electrical current—would register a change in its value.

The quantum tunneling of electrons becomes much more likely when the tip of the pin moves closer to an atom. A difference of one tenth of a billionth of a meter (also known as an angstrom) can change the electrical current by a factor of 10. Atoms themselves are at least a few angstroms in size, so as the tip of the pin passes over an atom, the current passing through the pin fluctuates upward dramatically, then falls again as the pin moves away from the atom. So by moving the tip of the pin over the surface of the gold foil, scanning it, they hoped to be able to draw a picture of the sea of gold atoms.

As improbable as it sounds, the experiment was a perfect success. Using a computer to translate the differences in electrical current into video signals, Binning and Roher saw what no one had ever seen before—an atom.

Four hundred years ago, Han Jansen and his son Zacharias invented the optical microscope: a matched set of lenses paired to magnify an image. By the middle part of the twentieth century, optical microscopes had reached their natural limits—the wavelength of light—and they *couldn't* get any better. But biologists and chemists wanted to peer into the tiniest details of the cell, which were themselves smaller than the wavelength of light (that is, smaller than one-half millionth of a meter). In the 1940s, physicists stepped in and created the electron microscope. Electrons can be used to create waves much smaller than the wavelength of light. This increased the power of the microscope enormously, but it still remained beyond the reach of even the most powerful electron microscopes to see a single atom.

The microscope that Binning and Roher had developed was at least a thousand times better than the best electron microscopes. It could create pictures of objects only a twenty-fifth as big as a single atom. The pair named their invention the scanning tunneling microscope, or STM.

The same electric charge that could be used to cause electrons to tunnel into solid matter could also be used to move that matter about. Soon experimenters using STMs discovered that the STM could bring order to the atomic world. One stunning example—from the IBM Almaden Research Center—was a photograph of a set of xenon atoms arranged to form IBM's logo! Suddenly it seemed as though the atomic universe—thought to be permanently hidden from view—had opened a window into its quirky world, a window that allowed us to reach in and touch the impossibly small. Chemists could now study the basic building blocks of the material world directly.

Within a few years, the innovation of the STM was made obsolete by the atomic force microscope, or AFM, which uses a laser beam to detect tiny changes in forces as its probe comes into contact with atoms. This method is much more sensitive even than the STM, and gives us an even finer picture of the atomic-scale world. And like the STM, the AFM can be used to shove atoms into new positions.

Although this may sound like child's play, the atoms seeming like so many children on a playground, it was actually quite difficult. Each experiment took many days or even weeks to prepare. Although the STM and AFM could read the surfaces of atoms, and although they proved invaluable to biologists and chemists around the world, they remained clumsy and very hard to use. But that was all about to change.

TOUCHING DOWN

At the start of the 1990s, it was generally acknowledged that the world's best virtual reality research facility existed at the University of North Carolina's Chapel Hill campus. Begun by Fred Brooks—another legendary figure in the annals of computer science—the program actually pre-dated the coining of the term *virtual reality*. Brooks, a pioneer of computer graphics, began his UNC career by putting graphics to work for chemists. He created tools that could ever more accurately model the behavior of molecules. Suddenly, chemists could see the results of their work in three dimensions, making it possible for them to theorize about how various molecules would interact. The entire field of pharmaceuticals became more of a science than an art. Chemists could now see how molecules could dock to one another, and from this, understand why drugs worked—or why they didn't.

By 1990, Brooks's colleagues were working on the design of Pixel Planes, which would become the fastest engine for computer graphics ever created. Instead of using a single computer, Pixel Planes divided its work among 250,000 separate processors. Each processor handled a part of the computation needed to create an image; the result was a dramatic improvement in the performance of computer graphics—even over Silicon Graphics's lauded Reality Engine.

One of the first uses of Pixel Planes was remarkably practical. UNC was planning to construct a new building to house its computer science program. Brooks got the architect's plans and had his graduate students create a simulation of the building long before it was built. Brooks and his students looked around inside the virtual building, even making a modification to the design by widening the balcony in the building's lobby. Because it had been simulated before it had been built, everyone got a better building. (This technique of simulating proposed designs has since become commonplace in architecture.)

In 1991, Warren Robinett—a video-game pioneer who had participated in the first virtual reality work at NASA Ames—was invited by Brooks to do research at UNC. Initially, he and Brooks thought that teleoperated robots could be a fruitful area of research, but two things blocked their path. First, robots were (and are) very expensive and difficult to maintain. Second, UNC had no program in robotics—they just didn't have the expertise they'd need to do interesting work.

Then Brooks thought back to his work with molecular modeling. The world of molecules was very real, and had brought him a good share of his renown. But Brooks had worked with simulated molecules, not with the real stuff of the universe, because molecules were too small to be manipulated by researchers. Until recently, that is. Robinett told Brooks about an old friend of his, Stan Williams, teaching at UCLA. Williams had tried to buy a scanning tunneling microscope for his program and found that they weren't for sale. So Williams and his graduate students *built* one.

Brooks already knew what an STM was. And now he could see where Robinett was going with this train of thought. The STM is like a robot, only it works on a very, very, very small scale. And the information that an STM sends out is inherently three-dimensional. It's the shape of atoms. If Robinett could figure out how to connect the STM to Pixel Planes, they'd be able to do more than create a flat video image; they'd be able to re-create the atomic world in VR.

Brooks assigned his star graduate student Russ Taylor to work with Robinett. Together they studied the mysteries of the atomic-scale universe. Stan Williams, happy to help, sent one of his graduate students to UNC, along with his homebrew STM.

After a few months of programming, the crew were ready for their first test. Williams had sent the graduate student out with some sample items that Williams had already imaged with his STM. These were graphite sheets (the carbon compound found in pencil lead) that had been bombarded with charged atoms. The atoms produced pits in the graphite, much as meteors do when they hit the surface of the Earth. The virtual world created by the combined STM/Pixel Planes system would—hopefully—look at least something like Williams's flat images from the STM.

When they flipped the machine on, what they saw stunned Williams's graduate student. The image before them looked very much like Williams's photos, but in three dimensions and with much greater detail. Already, the student could locate structures—atomic features—that hadn't been visible in Williams's photographs. By connecting the atomic world to the virtual world, Robinett and Taylor had dramatically improved their view of the surfaces of atoms. Atoms are three-dimensional; they take up space. A photograph can't reveal everything about them. But a virtual world, composed from the same information, is capable of revealing forms that the still photograph can't begin to represent. As impressive as that was, they had still another goal in mind.

In 1982, UNC had been the grateful recipient of an Argonne

Remote Manipulator, or ARM, an early haptic (touch) interface. They'd already connected it to their virtual reality systems, giving them the ability to create virtual worlds with haptic feedback. Now Taylor and Robinett integrated the ARM into their STM/Pixel Planes combo. Using the three-dimensional data from the STM, they built a system that would allow them to *feel* the surfaces of individual atoms. You could put your hand in the ARM, run it across the sample inside the STM, and feel the bumps, the rises and falls, as it passed across atoms.

In many ways, this was more significant than the original breakthrough of the STM/Pixel Planes combination. Suddenly chemists could *see* and *feel* individual molecules, something they'd never imagined they'd be able to do, providing a much more complete sense of the atomic-scale world. Our bodies are flooded with haptic information, and we can easily feel differences in materials that our eyes can't detect. When combined, sight and touch are a potent combination. The UNC team had succeeded in building a "nanomanipulator."

Chemists and biologists came to UNC to use the nanomanipulator. In just a few days, a scientist could learn more about a substance than he might in years of conventional study. A biologist could feel his way across the structure of a protein or caress a virus. To the delight of physicists, the nanomanipulator could push atoms around, but now they'd push back! The forces of the atomic world had been magnified and made tangible.

Today, the nanomanipulator project is one of the centerpieces of UNC's computer science program. Robinett left UNC, only to return a few years later, and Taylor, after working in the VR industry, returned to become one of the project leaders. They've added a few things, such as that lovely haptic interface from SensAble technologies that so captivated me in 1998, and they've made it available to the public. Late in 1999, a nearby high school was invited to conduct experiments on the nanomanipulator, communicating with it over the Internet. The nanomanipulator has sophisticated software that allows it to transmit its images and tactile information across a network, so, in theory, it could be used from anywhere on the planet.

Although the worlds of nanotechnology and virtual reality started out on distinct, separate paths, the nanomanipulator proves that they are, in fact, convergent trends. Scientists have wondered what virtual reality might be good for; they've also questioned how we'd ever be able to manipulate the nano-scale universe meaningfully. The nanomanipulator demonstrates that *together*, the two make perfect sense. The virtual world is most useful when it represents reality, and the nano-scale world makes most sense when it is represented virtually. Imagination and practicality are two sides of the same coin, neither complete without the other. All of our fantastic computer graphics have come to serve a useful purpose in the real world.

The software controlling the earliest nanomanipulator is freely available. One night, surfing the UNC website, I found detailed instructions on how to create my own interface to the nanomanipulator—should I want to. (A start-up company is now offering a nanomanipulator software package, so I could always go out and buy it.) Of course, a Pixel Planes isn't necessary anymore. In fact, the Sony PlayStation 2 should have enough graphics horsepower to bring the atomic-scale world into view in my living room. And the price of atomic force microscopes continue to drop. Although they cost about $75,000 today, in a few years they'll sell for just a few thousand—maybe even less.

On my ninth birthday, I was delighted to receive the gift of a microscope from my parents. I remember looking at slides of fleas, of amoebas, even pricking my finger and squeezing a few drops of blood onto a plate to get a glimpse of my own red blood cells. It opened the world of the very small to my view, made the mythical stories of cells suddenly very concrete, and helped me toward an enduring interest in the life sciences.

The next generation of children—those born today—might

receive the same optical microscopes, but these will be mere stepping-stones on the way toward a view of the world that has only just come into being. Perhaps they'll access these atomic-scale worlds from their PlayStations, connected across a high-speed planet-spanning Internet, learning about the raw stuff of the universe as I once wondered at my body's own blood cells. Or perhaps under their Christmas trees they'll find their own nanomanipulators, portals to the engines of creation.

EVERYTHING THAT RISES MUST CONVERGE

In 1924, Harvard University appointed Alfred North Whitehead to a professorship in its Philosophy Department. Although world-famous for his work in mathematics, Whitehead had never been considered a philosopher—at least, not one in the grand tradition of Aristotle, Immanuel Kant, or Friedrich Nietzsche. Instead of focusing on "the true, the good, and the beautiful"—the enduring themes of philosophy—Whitehead concentrated his intellectual efforts on the philosophy that lay *beneath* science.

This passion had led him to collaborate with mathematician Bertrand Russell (who would later win the Nobel Peace Prize for his role in creating the Campaign for Nuclear Disarmament and who gave us that ubiquitous icon of the 1960s, the peace symbol) on a text that both hoped would demonstrate the pure logic beneath mathematics. While mathematicians presumed that a deeper logic supported their work, no one had actually ever attempted to formalize the relationship between logic and math. Most assumed that it would be a relatively easy task. It wasn't.

After nearly ten years of effort, the pair published *Principia Mathematica* (Latin for "Principles of Mathematics"). Built up from nothing other than the rules of logic—the same rules computers apply at blinding speeds—it took them over *900* steps to get to "A = A," a proposition so obvious that a six-year-old child has no trouble mastering it. But no one had ever asked why "A = A," why we logically understand what makes one thing equivalent to another. The intuitions of mathematics, easily grasped, are light-years away from their logical expression.

The effort was so consuming—falling particularly hard on Russell's shoulders—that in later years he claimed the effort "broke him." (Indeed, Russell's contributions to mathematics came to an end with *Principia Mathematica*.) Despite this, both men saw the effort as immensely worthwhile. Their work clarified the logical relationships children take for granted when they perform addition.

After completing *Principia Mathematica*, Whitehead went on to become an influential educator, advocating reform in the instruction of mathematics, which previously consisted of the haphazard memorization of various principles and theorems. Nothing in the educational system drew all of the pieces of mathematics into a cohesive whole. Whitehead argued that students should be taught the logical foundations of mathematics; he believed that if a child had a firm grounding in the ideas that underlie mathematics, its advanced concepts would come more easily.

In 1980, as I entered MIT, I enrolled in Calculus with Theory. A full year of calculus is a requirement for all freshmen at MIT; as I'd already had calculus in high school, I was hungry for more than just a rehash of the same old formulas. What I got—and what Whitehead would have praised—was a development of calculus from the basic rules of mathematics (ten statements known as the Peano axioms, after their proponent, Italian mathematician Giuseppe Peano, who had a singular influence on both Whitehead and Russell). For me, calculus had been an abstract and nearly incomprehensible jumble of limits, derivatives, and integrals. Yet over the course of that semester, my intellectual fog lifted: calculus became entirely reasonable, logical, and sensible.

I'd leave my lectures with pages of notes, step-by-step logical arguments, which I'd then work through, following the same steps, adding to my knowledge as I used these arguments to prove new arguments. A house of understanding was built from the bricks of theory.

So much of what we know seems disconnected, distinct and isolated. Whitehead rejected this; he came to Harvard ready to talk about the deep relationship of all things, that *the act of knowing cannot be separated from how we come to know it*. We know what we know because of our experience with it; practice within the world gives us a more perfect understanding of it. In this, Whitehead declared himself to be against the mainstream of philosophy, which, since Plato, argued that philosophical truths were somehow outside the tide of human understanding, existing above it, perfect and complete. Practice makes perfect, Whitehead argued, and in 1929, he wrote *Process and Reality* to prove his point. Arguably the densest philosophical text of the twentieth century, it advanced the idea of the philosophy of organism, which states, at its essence, that the encounters between objects—including people—in space are the only real things; the vague and lofty ideals of Plato and Plato's intellectual children are nothing but wishful constructs.

Whitehead based his lessons on recent history. As he worked on *Process and Reality*, the science of classical physics, advanced by Isaac Newton a quarter of a millennium earlier and praised as objectively true by philosophers ever since, had utterly collapsed in the face of the still-new quantum physics. Whitehead surmised that this sudden collapse had less to do with science than with the dogmatic philosophical positions adopted by physicists, who felt they were defending the actual reality of the world as they fought off a steadily increasing body of evidence that seemed to indicate that Newton's depiction of the physical world was woefully incomplete. Although physicists have moved on and adopted the principles of quantum physics, most philosophers remain loyal to the classical error, searching for an objective standard of the true, the good, and the beautiful, a mirage Whitehead termed "misplaced concreteness."

Looking back on the history of artificial intelligence, we can see something of a redux of the battle between Whitehead's and Plato's heirs. The earliest pioneers of AI believed human intelligence to be something that existed as an objective reality and thus tried to program that reality into their machines. They failed. It took Rodney Brooks, with a robotic take on the philosophy of organism, to lead AI into a realm where real intelligence grows *because* of its encounters within the real world. In play we learn truth.

In an artwork like Char Davies's *Osmose*, Whitehead's arguments can be tested. If the relationship between objects in space teaches us what we know, then changing that relationship should produce a change in our knowledge. *Osmose* changes the body's relation to space, and in doing so, powerfully transfigures the participant, producing a new quality of being many could describe only as angelic. Such a transformation would be impossible in a world of platonic absolutes. Like the bits of evidence that brought Newton's theories down, we can see the twentieth century producing a series of experiments flying in the face of Plato's assumptions about the immutable laws of our existence.

Finally, there is Robinett and Taylor's nanomanipulator. It allows chemists and biologists to run a virtual finger across the surfaces that had previously been only theoretical. Suddenly, because the senses have been engaged, because the invisible has been made an observable body in space, scientists are capable of intuiting more truths than their theories could ever allow. The playful stroke of a tungsten pin over a graphite plate overturns theory in favor of experience, and provides the same sense-based foundation that underlies our understanding of the material world.

Using the nanomanipulator, more can be known with the hands—and understood—than can be learned in the lecture hall or from the pages of a textbook. The real world, when touched and played with, is the best tutor.

Whitehead advanced another idea with his philosophy of organism, that of *concrescence*, the production of novel togetherness. As things grow *together*, they become a new, unique thing. Elements that seem unrelated can through time emerge as a whole.

The history of the playful world is a history of concrescence; seemingly distinct advancements in artificial intelligence, robotics, virtual reality, and the Internet actually comprise one overarching wave of transformation, changing what we know by transforming how we come to know it. It is only now, after most of these transformations have played themselves out, that we can begin to see the outlines of the new thing sweeping them up and making them one.

PRIVATE LANGUAGES, PART III

There is no question that our children will be different from us; the only question is how different. We are different from our parents and our grandparents, insofar as the way we came to learn of the world around us differed from their experiences. We do not possess the simple faith of medieval peasants, with their angelic cosmologies and saintly hagiographies, nor can we embrace the idea of the grand watchmaker, a view of the universe expressed by Isaac Newton's contemporaries. (If we did, our computers, powered by quantum physics, would become incomprehensible to us.) Each age has a truth, an explanation appropriate to itself and its relation to the world.

These truths, although they can be embodied in cultures, do not develop from the cultures themselves. As they grow into awareness, children learn the truth of the world from their experience in the world. Cultures create toys to instruct children in these truths, and through these toys children learn the possibilities and boundaries of the real.

We are very quick, these days, to translate our shifting view of truth into artifacts for our children. The Furby embodies some of our latest intuitions on the nature of intelligence—and that's only the first in what will be an ever-accelerating evolution of conscious devices straddling a middle ground between the quick and the dead. Already, toy designers are working on another generation of dolls that will learn from and react to their caretakers, using principles pioneered by Rodney Brooks and his team at MIT. This reflective intelligence will become a persistent feature of the world our children are entering. After it makes its way into dolls, it will find its way throughout our world, becoming pervasive. The dead artifacts of culture will listen to us, react to us, and learn from us, becoming more and more alive.

In fewer than three generations, the Lego has gone from a children's building block to an archetype that has come to define the way we will soon manipulate the essential elements of the physical world. Yesterday's child slapped bricks together to create a miniature-scale model of a house or a boat; tomorrow's child will place atom against atom. As we grew up, we could not see these atomic forms; for us, they remained theoretical and invisible. But now anyone can reach out and touch an atom, or push some around to build a molecule. This grants atoms a material reality that allows us to think of them in the same concrete terms as we might think of a stack of Legos.

When the active intelligence of the computer is added to the mix, the hard-and-fast rules of the material world—the notion that objects are solid, static, and consistent—vanish. We are on the threshold of an era of active objects, designed from the ground up, atom by atom. They will contain their own intelligence, so they can listen to us and respond to our needs. Although solid, they will be changeable, able to transform themselves nearly instantaneously, to meet the requirements of the moment.

Those who have grown up in a world where the physical persistence of objects is an assumed absolute might find the plastic reality of ever-mutable objects to be a disorienting nightmare; our children will know nothing but the possibility of change. They will accept that physical objects, as solid as they might appear, are just a momentary configuration, a fleeting thought that can be made and unmade according to the heart's desire.

Even as the universe becomes more flexible, we find that our presence within it, no longer fixed at a single point, can move seamlessly throughout the collected knowledge of humanity. We can place our eyes and hands into other worlds—distant lands, synthetic or impossibly small—and each site we visit will change how we think about the world. Virtual reality is the imagination realized, the hidden parts of ourselves brought into view. The World Wide Web is a ubiquitous fabric of knowledge, allowing us to know nearly anything, at any time we might desire, anywhere we might need to know it. We have not freed ourselves from our bodies—far from it—but, instead, the physical divisions between this and that, near and far, erode in the face of technologies that put us here *and* there, let us touch a single atom, and allow us to gaze upon the entirety of the Earth. Our children will come to apprehend a different reality than the one we inhabit. Because we have bestowed upon them eyes and ears and hands we could only dream of, they will be granted a broader sense of self.

Nearly a hundred years ago, Jean Piaget watched his children as they practiced philosophy with greater intensity than any philosopher, putting their self-developed logical propositions on the nature of reality to the test, amending their views as the errors of their ways became clear, using one discovery as a platform for the next, building their understanding of the world from their experience within it. Our children—and all children, for as long as humanity exists—will do precisely the same thing. But the rules—or rather, our understanding of the rules—have changed utterly. That which was impossible for our grandparents is now commonplace; that which seems impossible to us will be matter-of-fact for our children.

All of this means that our children are engaged in developing a new philosophy, which will articulate the truths of a new age. Every bit as human as ourselves, they will nonetheless be forced to develop ideas and understandings that will seem alien to us, important tools they will need to make sense of the playful world.

Will we fear this change?

THE PLAYFUL WORLD

A half a century ago, anthropologist Margaret Mead noted that the pace of change in the Western world had accelerated to the breaking point. In earlier, slower times, an older generation could educate the young in the wisdom of their ways—important lessons that would serve the young well. As cultural development began to accelerate, this generational transmission of values became progressively less useful, finally turning into a generation gap. The younger generation had come to believe—with some justification—that its elders had little of real value to teach them, just the obsolete experiences of a bygone era.

The elders of that generation reacted with fear, branding youth with a series of evolving nicknames and resisting the cultural truths that members of the younger generation created for themselves without the benefit of the wisdom of its elders. In the Western world, the generation gap erupted into the fissure of the 1960s, when youth struck out against the elder generation, seeking to deprive them of their authority over the young.

As a civilization, we have not come to terms with this fissure. Sociologists and pundits sought explanations for it in the changing role of mass media or in the redefinition of family and work, but despite some serious efforts, no real understanding of that schism has taken root in our cultural consciousness. And those who fail to learn from history are condemned to repeat it.

Such a situation greets us at the dawn of the third millennium.

As the children of the playful world grow, as they internalize their new truths and translate them into new cultural values, an older generation is likely to react with shock and horror as they see the truths of their own culture ignored (or defied) by a generation who knows that the rules have changed, who have learned a different way of being in the world. Yet these children will have the benefit of experience that we lack, and from a very early age they will understand things very differently than we do. Their language and customs may be strange to us—but not because they are bad or immoral; they simply reflect a deeper understanding of the world we have created for them.

If history is a guide—and I hope it is not—we will react poorly to the emergence of this new truth and the culture that necessarily travels with it. Cultural historian William Irwin Thomson has remarked that the forms of a new culture are always seen as demonic by the culture that precedes it. If this is true, then we may see the new truth as diabolical or inhuman and try to snuff it out. We would be making the same mistake that another generation made half a century ago. And, while the possessors of this new truth might simply sit back and wait for us to fade away (though this seems unlikely), we would be doing an even greater disservice to ourselves. For we have given birth to our teachers. Our children will know how to make sense of the playful world, an important lesson they will be happy to share with us, if we are willing. Reversing the flow of history, if only momentarily, we will need to learn how to speak the language of this new world, its customs and truths.

It might be humbling (after all, they are our children), but in that humility is a great opportunity: we could resist, to be shoved aside by history, or we could choose another path, listen to our children intently, and let them teach us their secrets, their new philosophies, and from this, grow into a new understanding of the world we have made for ourselves.

This, I believe, represents an appropriate way to manage the rapid transformations of the early twenty-first century. Our children, already fully involved in a world we have yet to recognize as our own creation, hold the keys to a future where everything that was solid will have faded into mist. We have to free ourselves of our own illusions about the way things work and make room for a new truth. There is a real danger here: if we fail to listen to our own children, how can we expect them to listen to us when we try to teach them of older, but still essential human values? In an age of ever-accelerating change, we must find a middle path, one that allows us to adapt to our evolving understanding of the world while maintaining our humanity.

Just a few months before he passed away from brain cancer, anthropologist and philosopher Terence McKenna was asked about this dilemma in an interview with author Erik Davis. He answered eloquently:

> We can build a civilization like nothing the world has ever seen. But can it be a *human* civilization? Can it actually honor human values? . . . the rate of invention or gross national product or production of industrial capacity—all of these things are all very well. But the real dilemma for human beings is how to build a compassionate human civilization. The means to do it come into our ken at the same rate as all these tools which betray it. And if we betray our humanness in the pursuit of civilization, then the dialogue has become mad.
>
> So it is a kind of individual challenge for every single person to demand that compassionate civilization.

We could allow technology to send us to the sidelines; if we let that happen, our generation will bear the responsibility for all of the horrors of a world massively transformed by pervasive reactive intelligence, widespread nanotechnology, and ubiquitous presence, because we chose to be rigid when we needed to be open-minded. If we decide instead to listen to the young and learn their lessons, we may find that our children, who have grown up surrounded by the magical toys we have given them, are making for themselves a world of play.

ACKNOWLEDGMENTS

The journey of *The Playful World* has been a short and happy one. Keith Kahla, an old friend, deserves thanks for hooking me up with my agent, Ira Silverberg. Ira was absolutely essential in helping me tune my ideas, turning them into a book proposal. He also managed the miraculous feat of selling this book. When I first spoke on the telephone with my editor, Peter Borland (*il miglior fabbro*), I knew I'd found someone who understood my thesis. I've received nothing but exceptional support from him and the entire staff at Ballantine Books.

The interviewees—Ken Goldberg, Mitchel Resnick, Tim Berners-Lee, Ted Nelson, Anne Foerst, Ralph Merkle, Joachim Sauter, Gerd Grüneis, Pavel Meyer, Axel Schmidt, Tony Parisi, Servan Keondjian, Warren Robinett, and Char Davies—all gave graciously of their time and made the additional effort of reviewing the draft copy of this manuscript. I must especially thank Philippe van Nedervelde for his extensive and thoughtful consideration of the sections on nanotechnology, and Rachel Chalmers for her forthright criticism of the first draft. My readers—Scott Moore, Greg Jacobson, Owen Rowley, and my mother, Judith Quandt—gave me feedback and encouragement during the writing process; my gratitude for their interest and patience cannot be measured. Erik Davis gave me some pointers that I used in "Intelligence" and allowed me to use a quote from his last interview with Terence McKenna.

Parts of this book were written in Florida, at the home of Ed and Judith Quandt, on Hawai'i, at the home of Serge Tretiakov and Masha Khazin, and at the home of Tony Parisi and Marina Berlin in San Francisco. I want to thank them for their hospitality—and their network access!

I alone am responsible for any errors that may have crept into the text.

INDEX